Proceedings of the 2009 Viral Clearance Symposium

• • • • • • • • • • • • • • • •

Developments in Biologicals

Vol. 133

This series "Developments in Biologicals" begins with Vol.102 and is the continuation of IABS symposia series "Progress in Immunobiological Standardization", Vols 1-5, "Immunobiological Standardization", Vols 1-22 and "Developments in Biological Standardization", Vols 23-101.

Organized and published by the International Association for Biologicals (IABS)

Basel · Freiburg · Paris · London · New York
Bangalore · Bangkok · Singapore · Tokyo · Sydney

Proceedings of the 2009 Viral Clearance Symposium

Indianapolis, IN, USA
20-21 March 2009

Editors

George Miesegaes
CDER/FDA

Mark Bailey
Eli Lilly and Company

Hannelore Willkommen
RBS Consulting

Qi Chen
Genentech, Inc.

David Roush
Merck Research Laboratories

Johannes Blümel
Paul Ehrlich Institut

Kurt Brorson
CDER/FDA

Proceedings of an International Symposium organized by:
The 2009 Viral Clearance Symposium Organizing Committee

36 figures and 38 tables, 2010

Basel · Freiburg · Paris · London · New York
Bangalore · Bangkok · Singapore · Tokyo · Sydney

• • • • • • • • • • • • • • • • •

Developments in Biologicals

Library of Congress Cataloging-in-Publication Data

Viral Clearance Symposium (2009 : Indianapolis, Ind.)
Proceedings of the 2009 Viral Clearance Symposium, Indianapolis, IN, USA, 20-21 March 2009 / Editors, George Miesegaes, CDER/FDA, Mark Bailey, Eli Lilly and Company, Hannelore Willkommen, RBS Consulting, Qi Chen, Genentech, Inc., David Roush, Merck Research Laboratories, Johannes Blümel, Paul Ehrlich Institut, Kurt Brorson, CDER/FDA.
p. ; cm. -- (Developments in biologicals ; vol. 133)
Includes bibliographical references.
ISBN 978-3-8055-9682-4 (soft cover : alk. paper)
1. Pharmaceutical biotechnology--Congresses. 2. Viral contamination--Congresses. I. Miesegaes, George, editor. II. International Association for Biologicals, issuing body. III. Title. IV. Series: Developments in biologicals ; v. 133. 1424-6074
[DNLM: 1. Drug Contamination--prevention & control--Congresses. 2. Viruses--Congresses. 3. Drug Industry--methods--Congresses. W1 DE997VKF v.133 2009 / WA 730]
RS380.V57 2009
615'.19--dc22

2010049427

Bibliographic Indices. This publication is listed in bibliographic services, including Current Contents® and Index Medicus.

Drug dosage. The authors and the publisher have exerted every effort to ensure that drug selection and dosage set forth in this text are in accord with current recommendations and practice at the time of publication. However, in view of ongoing research, changes in government regulations, and the constant flow of information relating to drug therapy and drug reactions, the reader is urged to check the package insert for each drug for any change in indications and dosage and for added warnings and precautions. This is particularly important when the recommended agent is a new and/or infrequently employed drug.

Worldwide distribution by S. Karger AG, Basel (Switzerland).
Printed in Switzerland by Instaprint.
ISBN 978-3-8055-9682-4

Opinions expressed in this document are those of the authors and do not necessarily represent the official views of the US government, individual firms or regulatory authorities.

Organizing Committee

George Miesegaes, CDER/FDA
Kurt Brorson, CDER/FDA
Qi Chen, Genentech, Inc.
David Roush, Merck Research Laboratories
Hannelore Willkommen, RBS Consulting
Mark Bailey, Eli Lilly and Company

Session Chairs

Kurt Brorson, CDER/FDA – Protein A Chromatography
Qi Chen, Genentech, Inc. – Low pH Treatment
David Roush, Merck Research Laboratories – Anion Exchange Chromatography
Hannelore Willkommen, RBS Consulting – Alternative Chromatographic Methods
Mark Bailey, Eli Lilly and Company – Virus Filtration

Symposium Participants

George Miesegaes – CDER/FDA
Mark Bailey – Eli Lilly and Company
Kurt Brorson – CDER/FDA
Qi Chen – Genentech, Inc.
David Roush – Merck Research Laboratories
Hannelore Willkommen – RBS Consulting
Johannes Blümel – Paul Ehrlich Institut
Pedro Alfonso – Centocor Ortho Biotech
Dayue Chen – Eli Lilly and Company
Parviz Shamlou – Eli Lilly and Company
Patricia Alred – Centocor Ortho Biotech
Rachel Specht – Genentech, Inc.
Glen Bolton – Wyeth
Eva Gefroh – Amgen
David Farb – MacroGenics
Michael Barry – ImClone Systems
Scott Lute – CDER/FDA
Olga Galperina – Human Genome Sciences
Brian Hubbard – Amgen
Lisa Connell-Crowley – Amgen
Markus Heitzmann – Novartis
Paul Mensah – Pfizer
John Mattila – Regeneron
Edwin Lundell – Abbott Bioresearch Center
Richard Wright – Wyeth
M. Ellen Dahlgren – Merck Research Laboratories
Franz Nothelfer – Boehringer Ingelheim Pharma GmbH
Mark Gustafson – Pfizer
Chris Dowd – Genentech, Inc.

Contributors

Nihal Tugcu – Merck Research Laboratories
Viveka Raol, Houman Dehghani, Tom McNerney – Amgen
Kris Barnthouse, Charles Felice – Centocor Ortho Biotech
Helmut Winter, Dorothee Ambrosius – Boehringer Ingelheim Pharma GmbH
Yaling Wu, David Kahn, Yuling Li – Human Genome Sciences
Lenore Norling, Sherrie Curtis, Bin Yang – Genentech, Inc.

Acknowledgements

The authors thank Kathryn King, Jack Ragheb, and Scott Lute (CDER/FDA) for careful review of this manuscript. This work was funded in part by a Cooperative Research and Development Agreement (CRADA) between CDER/FDA and Eli Lilly and Company.

Abbreviations

AEX:	Anion exchange chromatography
CEX:	Cation exchange chromatography
HA:	Hydroxyapatite
HIC:	Hydrophobic interaction chromatography
CTO:	Contract testing organization
X-MuLV:	Xenotropic murine leukemia virus
MMV:	Minute Virus of Mice
LRV:	Log_{10} reduction value
CHO:	Chinese Hamster Ovary
BE:	Bind and Elute
FT:	Flow through
S/D:	Solvent detergent
FDA:	Food and Drug Administration (United States)
PEI:	Paul Ehrlich Institute (Germany)
Q-PCR:	Quantitative polymerase chain reaction
HCCF:	Harvested cell culture fluid
IND:	Investigational new drug
BLA:	Biologics license application
IW:	Intermediate Wash
WTB:	Wash to baseline
Polio:	Poliovirus (usually Type 1 in validation studies)
ERV:	Ecotropic Retrovirus
Reo-3:	Reovirus type 3
HCP:	Host cell proteins
A-MuLV:	Amphotrophic murine leukemia virus
E-MuLV:	Ecotrophic murine leukemia virus
BDFP:	Binding domain Fc fusion protein
HMW:	High molecular weight
LMW:	Low molecular weight
POI:	Product of interest
RVLP:	Retrovirus-like particle
BVDV:	Bovine viral diarrhea virus
DoE:	Design of Experiments
QbD:	Quality by design
QSFF:	Q-sepharose fast flow
IT:	Integrity tests
FP or RFP:	(Receptor) Fc fusion protein
PRV:	Pseudorabiesvirus
PPV:	Porcine Parvovirus
SV40:	Simian virus 40
mAb:	Monoclonal antibody

Definitions

Complete clearance: Complete clearance of virus by a unit operation (i.e. no virus detected in the output process fluid in a spiking study). For the purposes of this report, an LRV with the presence of a ">" or "≥" sign indicates complete clearance.

Incomplete Clearance: Incomplete clearance of virus by a unit operation (i.e. residual virus is present in the output process fluid in a spiking study) or a unit operation where complete clearance is not claimed. For the purposes of this report, an LRV with the absence of a ">" or "≥" sign indicates incomplete clearance.

Effective clearance: Depending on the virus and unit operation, 4 $\log_{10}$ or greater of viral clearance is considered to be "effective clearance"

Robust clearance: Viral clearance that is generally consistent within ranges of operating parameters, when employed under typical manufacturing conditions.

Contents

Miesegaes G, Bailey M, Willkommen H, Chen Q, Roush D, Blümel J, Brorson K (eds): Proceedings of the 2009 Viral Clearance Symposium. Dev Biol (Basel). Basel, Karger, 2010, vol 133, p 1.

SESSION I

Overview

Miesegaes G, Bailey M, Willkommen H, Chen Q, Roush D, Blümel J, Brorson K (eds): Proceedings of the 2009 Viral Clearance Symposium. Dev Biol (Basel). Basel, Karger, 2010, vol 133, pp 3-6.

Overview

Conference Scope and Purpose

The 2009 Viral Clearance Symposium (Indianapolis IN, USA) was held to interactively discuss methods for virus removal and inactivation during biopharmaceutical manufacture. Its genesis was the result of worldwide regulatory and industry recognition that challenges, gaps, and opportunities for improvement exist, that if formally addressed could benefit the field as a whole. Participants included regulatory speakers from the US and EU, as well as speakers from a number of mid-size to larger companies.

Representatives from a total of seventeen firms, from the US, Europe, and Japan had participated, most giving short presentations of their firm's experience with viral clearance. They were invited on the basis of possessing sufficient experience in the area of viral clearance, and the ability to informally discuss their data sets. At the time, all participating firms at the time had at least one marketed biotech product and/or a substantial number (i.e. >3-4) in early (IND) phases.

The symposium was organized into five sessions, each targeting a specific class of biotech unit operations typically employed in viral clearance studies. These included affinity chromatography, acid pH inactivation, ion exchange chromatography (including membrane adsorbers), detergent-based inactivation, and virus filtration. Additionally, new or uncommon chromatography operations, having limited industry-wide experience, were grouped into an "alternative chromatography methods" section. Each session consisted of 5-12 presentations on a range of topics including individual case studies, overviews of the firms' overall experience, "lessons learned"-type scenarios, and quality by design (QbD) approaches to viral clearance. Based on a group assessment of the data presentations, each session resulted in a work product that included (1) a list of conclusions from the session (i.e. "What we learned"); and (2) recommendations for further actions (i.e. "What are the next steps").

The symposium started with a presentation by FDA (USA) of a large-scale analysis of data from regulatory submissions over the past twenty years. A similar presentation (involving data mining of regulatory submissions) from Paul Ehrlich Institute (PEI, Germany) followed. In these two presentations, and a subsequent series of brief industry presentations covering various unit operations, it was made clear that many unit operations are quite effective in clearing viruses. This was particularly true of low pH inactivation (albeit for certain viruses), anion exchange chromatography, and virus filtration. Moreover, the follow-up discussions at the end of each session, and the wrap-up at the end of the symposium, aimed to synthesize the regulatory data mining knowledgebase with the industry-generated data. The symposium also revealed a number of unknowns in the field which were defined and prioritized, and in closing served as potential action items for future experimental studies.

Technological Background of Viral Clearance in Biotechnology

Bioprocessing consists of sequential unit operations that capture and/or purify cell culture derived protein products. These operations, or steps, can vary significantly in mechanism of action and in many cases focus primarily on removal of product and process impurities while maintaining acceptable yield (i.e. serendipitously, where viral clearance occurs as a secondary outcome). Protein A chromatography, for instance, is the most common technology employed to capture cell culture derived antibody products. This technique utilizes interaction domains within the Protein A molecule, which bind the antibody Fc region with relatively high affinity and specificity. Of concern however is the heterogeneous nature of the HCCF feedstock, which makes this a complex unit operation to fully understand from a mechanistic standpoint. Other chromatographic techniques such as ion exchange (e.g. AEX and CEX chromatography) are commonly employed downstream, thus serving as polishing steps. Examples of non-chromatographic unit operations include various forms of chemical virus inactivation and virus filtration. These latter operations are also examples of dedicated or deliberate clearance steps, where virus clearance is the primary impetus for their utilization. All of the above technologies have been successfully employed for clearing a number of viruses of both known and potential concern [1,2].

Regulatory Agency Experience of Viral Clearance

Given the large number of industry products in development, government agencies have indirectly collected a substantial amount of viral clearance information, i.e. as attained from within various regulatory submission documents. Unfortunately however, this information is not typically received in a pre-sorted or complete format. Therefore, a significant amount of data mining, compilation and sorting would be required in order to place it in a format more suitable for analysis (e.g. identification of trends, making definitive conclusions concerning aspects such as robustness, etc.). Nevertheless, given the breadth of potentially useful information, both the FDA and PEI have initiated programs to mine their regulatory databases specifically for these purposes.

US Food and Drug Administration (FDA) – Generated database of viral clearance *(G. Miesegaes)*

The FDA viral clearance data mining project was initiated in the summer of 2008. Viral clearance information from over 200 regulatory submissions spanning the past twenty years (IND, NDA, MF, and BLA document types) was extracted and assimilated into a database. Product information, manufacturing process conditions, and viral clearance data (expressed as LRV) were collected to the extent feasible and subsequently analyzed for trends. The scope of the database was limited to products reviewed by CDER's Division of Monoclonal Antibodies (DMA). Most of the product base consisted of full-length, mammalian cell culture generated, unmodified antibodies of the human IgG_1 subtype (Fig. 1). Information submitted for antibody-drug conjugates and fusion proteins was also collected, as many are reviewed by DMA.

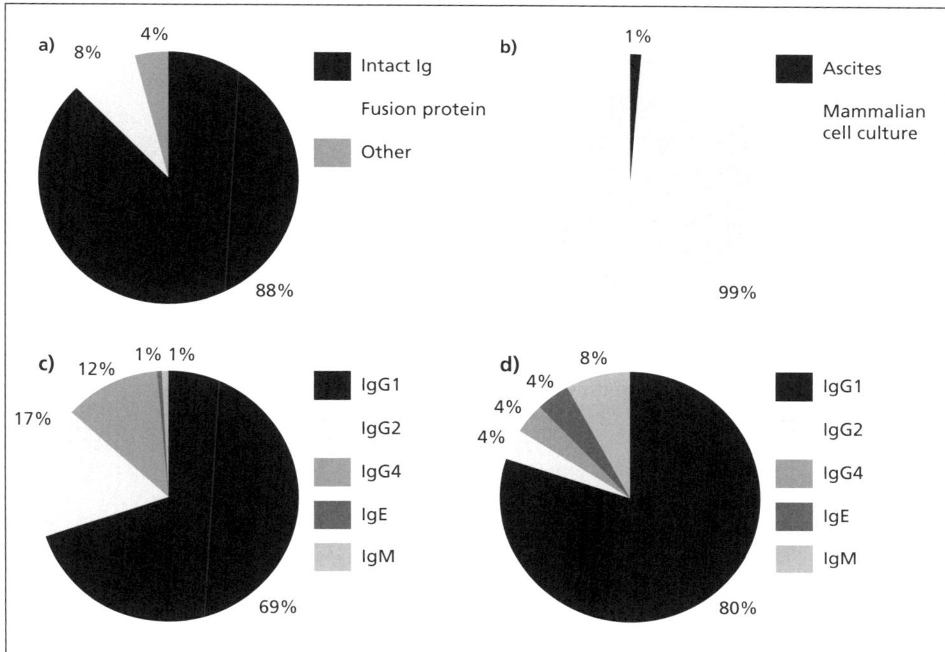

Fig. 1: Summary of antibody products in CDER's viral clearance database. (a) Percentage by product type: Intact Ig, fusion protein, other (i.e., Fab, Fv, etc.). (b) Percentages of ascites and cell culture derived antibody products. (c-d) Percent of Ig subtypes for intact (c) and fusion/other (d) antibody products. See text for detail. n=205. Data courtesy of George Miesegaes, FDA.

In this report, a "study" consists of all viral clearance information submitted in a single regulatory document for a particular unit operation, whereas a "record" comprises data from a single study for a given virus under specific parameter settings. Under this system a given study may contain information sufficient to generate multiple records. This was an important consideration during the database design phase as it was critical to maintain the ability to isolate and assess distinct experimental conditions; process parameters and other variables (including virus species assessed), especially in QbD studies, may yield a wide range of clearance values.

Ultimately, the goals of the database project were three-fold:

1) To centralize and index the numerous viral clearance studies into a single repository. This served not only to enhance record-keeping, but also allowed for a series of internally-performed statistical-based comparisons [3];

2) To synthesize the data and establish a knowledgebase to identify key strengths and gaps in current practices. Aspects of this included the identification of effective virus clearance steps and the determining of characteristics that might associate with outliers/ failed studies;

3) To define, prioritize, and subsequently perform hypothesis-driven verification experiments.

At the beginning of the Symposium, FDA presented pie chart distributions of the percentages of records reporting particular LRVs for various unit operations. These included:

a) Protein A, AEX, and CEX chromatography;

b) Low pH, S/D, and detergent-only inactivation steps; and

c) Virus retentive filtration using large- and small-pore size filters.

This information was presented in order to provide a snap-shot of industry experience regarding the robustness and effectiveness of each of these unit operation classes, and is discussed below.

Analysis of the database shows that in general, many of the current technologies employed to remove or inactivate viruses are effective and robust. As one example, filters specifically designed to retain viruses generally achieve several log_{10} LRV with typical model viruses such as MuLV (5-7 log_{10}) and MMV (3-7 log_{10}) (refer to Session VII), as well as other viruses (data not shown). For the Symposium, FDA focused its presentation on two viruses: MuLV, a large diameter (>80 nm) retrovirus, and MMV, a small diameter (18-26 nm) non-enveloped parvovirus. Both models are commonly studied to assess viral clearance [4,5].

It was found that some unit operations (e.g. Protein A chromatography) are perhaps not as consistent as others (e.g. AEX) for viral clearance. As might be expected, these tended to be operationally more complex compared to others (e.g. BE mode columns tend to use more buffers). In essence it became apparent that additional work would be required in order to better understand these more complicated operations. Additional findings from the FDA database included:

- Infectivity and Q-PCR detection methods can produce different LRV for a given unit operation.
- LRV can vary even within the same unit operation class and be reported either as "complete" (i.e. as "> x log_{10}"), or in other cases, a claim of complete clearance is not made.

Paul Ehrlich Institute (PEI) - Generated database of viral clearance *(J. Blümel)*

A similar data analysis project was performed considering clinical trial applications for 91 different monoclonal antibodies received at PEI between October 2004 and February 2009. The analysis focused on two methods for virus reduction: (1) Inactivation of retroviruses at low pH incubation; and (2) Virus retentive filtration of X-MuLV and parvoviruses.

Miesegaes G, Bailey M, Willkommen H, Chen Q, Roush D, Blümel J, Brorson K (eds): Proceedings of the 2009 Viral Clearance Symposium. Dev Biol (Basel). Basel, Karger, 2010, vol 133, p 7.

SESSION II

Protein A Chromatography

Miesegaes G, Bailey M, Willkommen H, Chen Q, Roush D, Blümel J, Brorson K (eds): Proceedings of the 2009 Viral Clearance Symposium. Dev Biol (Basel). Basel, Karger, 2010, vol 133, pp 9-22.

Protein A Chromatography

Technological Background

Protein A chromatography is the most commonly employed technology for the capture of cell culture derived antibody products. Protein A efficiently separates antibodies and Fc fusion proteins from complex cell culture harvests via a specific interaction with the antibody Fc region. The heterogeneous nature of the HCCF feedstock makes this a complex unit operation. The commonly understood mechanism for viral clearance by Protein A is that viruses largely flow through with most non-antibody cell culture components during the loading phase and post-load wash. This is not a complete process, as a few viruses presumably bind to the media by non-specific electrostatic (or other) interactions. These viruses are then presumably liberated by the pH transition associated with elution, leading to variable levels of virus in the eluate. Because this process does not involve precise molecular interactions or controllable parameters, it can lead to variable LRVs when chromatographic separation power is assessed alone by PCR based methods.

A second mechanism of clearance is by inactivating viruses via the low pH of the elution buffer. This is effective only against susceptible virus (generally enveloped virus) and is dependent on exposure time. The exposure time can be variable depending on column buffer, linear flow rate, and neutralization strategy (i.e. the decision to neutralize immediately or hold at low pH for a period of time). This variability is why most firms include a discrete low pH inactivation step that is studied separately by infectivity assays.

Regulatory Agency Presentation

FDA *(G. Miesegaes)*

Figure 2 shows the mean LRV and the ranges of clearance achieved by Protein A for two viruses, MuLV and MMV. We separated infectivity assay data from PCR-based assay data (Fig. 2a, c-f), as well as whether the Protein A step was serving as a capture vs. a downstream processing step (Fig. 2b). We found that in the case of the enveloped retrovirus MuLV, clearance measured by Q-PCR was significantly lower than for clearance measured using infectivity assays (Fig. 2a, $p<0.0001$). This finding supports the notion that infectivity assays measure viral inactivation from the low pH elution buffer in addition to the separation power of the column. This additional component seems to add on average a 1.7 $\log_{10}$ clearance power to this step. In contrast, for MMV, which is non-enveloped and typically not inactivated by low pH, we found little difference in mean LRV when measured by infectivity vs. Q-PCR.

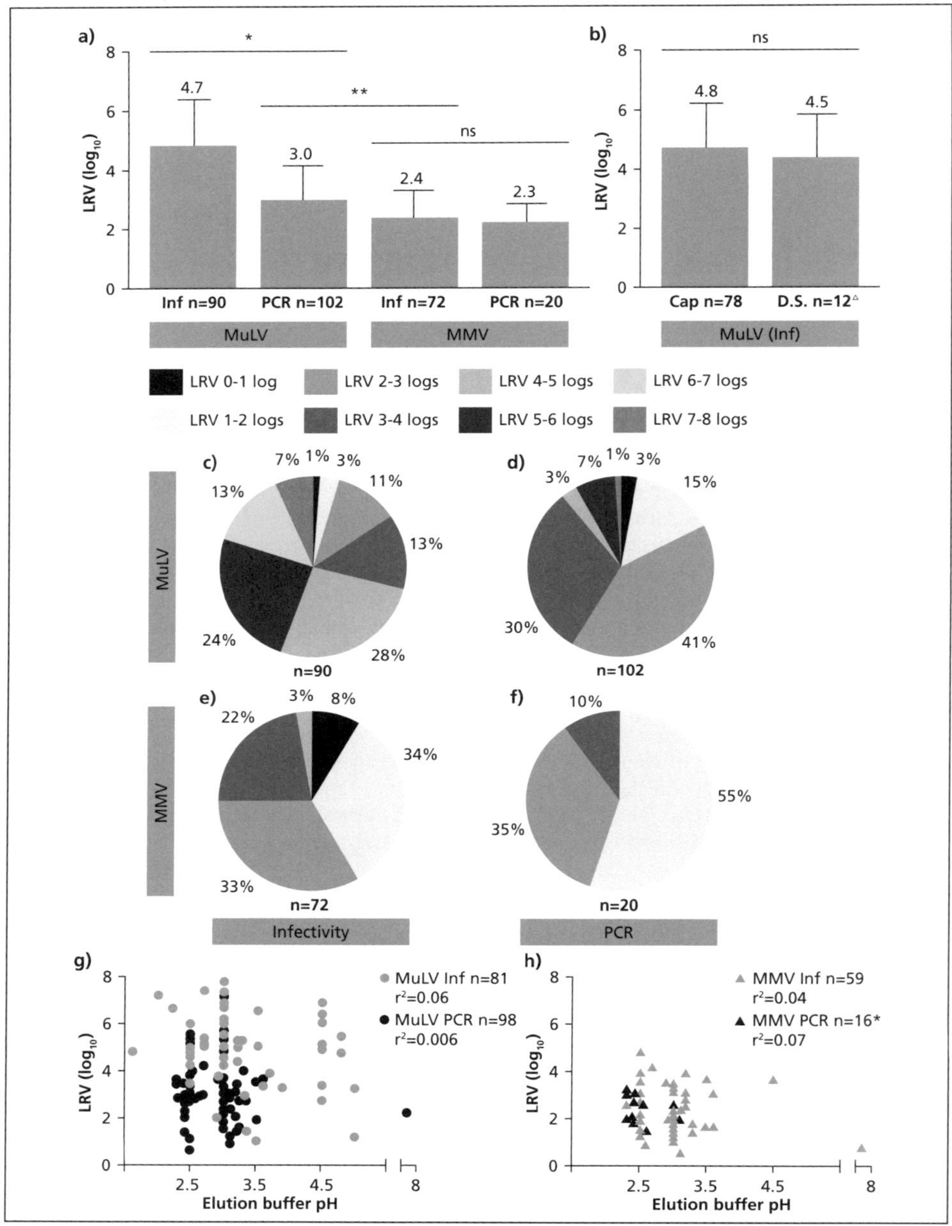

Fig. 2: LRV identified for Protein A chromatography in CDER's viral clearance database. (a) Comparison of infectivity (Inf) to PCR based detection for MuLV and MMV (+/- standard deviation). (b) Protein A used as a downstream step vs. as capture (MuLV shown). (c-f) LRV ranges identified for MuLV (c-d) and MMV (e-f). Pie charts show MuLV records binned into 1 $\log_{10}$ LRV ranges. Values (refer to key) are expressed as a percentage of the total number of database records in the particular 1 $\log_{10}$ bin. (g-h) Scatter plot of elution buffer pH and LRV. *$p<0.0001$; **$p<0.0002$; ns not significant; $^{\Delta}$low number of entries noted. Data courtesy of George Miesegaes, FDA.

Interestingly, compared to MMV (infectivity vs. Q-PCR), the mean Q-PCR-measured MuLV LRV was somewhat greater (3.0 $\log_{10}$ vs. 2.3 $\log_{10}$; p=0.0002). Thus MMV may be more challenging to remove via Protein A chromatography, for reasons beyond low pH inactivation.

Capture step feedstocks are complex mixtures of cell culture and media components as well as the product of interest. In contrast, downstream polishing steps process feedstocks that are relatively free of cellular debris, HCP and DNA. To determine if feedstock complexity might impact Protein A viral clearance, we compared MuLV clearance data when Protein A was the first step in the purification train with data for Protein A positioned further downstream (Fig. 2b). The assessment did not reveal a substantial difference, implying that viral clearance by Protein A chromatography is not overtly affected by location in the process or feedstock complexity. However, differences in other unit operations were observed (e.g. Session IV). It is worth noting that data records containing LRV < 2 did, however, use complex feedstocks (either ascites or a high-titer product) as their source [3]. Lastly, data presented by Amgen (Session II) also suggest an influence of feedstock on viral clearance by Protein A.

Clearance assessed by PCR-based assays is considered more precise than clearance measured by infectivity assays [6]. For Protein A, Q-PCR exclusively measures the separation power of the column. Therefore, to more accurately compare the range of MuLV and MMV LRVs, all PCR records were separated out from infectivity assay records (Figs 2c-f). We made the following observations:

- Based on mean values (Figs 2a, b), the PCR-measured MuLV LRV distribution was skewed consistently lower than those measured using infectivity assays. This was not seen for the non-enveloped and low pH-insensitive MMV:
  - Only 1% of records using PCR found more than 6 $\log_{10}$ of clearance for MuLV.
  - 86% infectivity and 41% PCR data reported the removal of 3 $\log_{10}$ or better for MuLV (Figs 2c, d)
  - Only 25% infectivity and 10% PCR data reported the removal of 3 logs_{10} or better for MMV (Figs 2e, f)
- There were more "complete clearance" vs. "incomplete" MuLV records; this was regardless of whether using infectivity or PCR-based assays (data not shown).
  - For MuLV Infectivity: n=51 incomplete and n=39 complete clearance values
  - For MuLV PCR: n=84 incomplete and n=18 complete clearance values
- No records reporting complete clearance of MMV by Protein A were found in the dataset.

Because it was noted that low pH elution seems to confer an extra 1.7 $\log_{10}$ of MuLV clearance power to Protein A, a correlation between elution buffer pH and LRV was examined in a scatterplot for all Protein A records, separated by virus (MuLV and MMV) (Fig. 2g, h). Surprisingly, no significant correlation ($r^2<0.1$ in all cases) was apparent; a finding that was also observed by an industry participant, albeit in a more restricted pH range (Session II). In FDA's case, this may have been a result of the limitations when performing retrospective database meta-analyses (heterogeneity of firms, contract testing labs, etc.) rather than the ineffectiveness of low pH. In addition, impacts of other process factors like feedstock complexity and intermediate washes (IW) can contribute also to LRV, but cannot be systematically evaluated in a retrospective analysis.

Overall, the database analysis shows that although Protein A chromatography may not be the most predictable removal step, it is capable of removing modest levels of both retroviruses (i.e., MuLV) and parvoviruses (i.e, MMV) when employed as either a capture or a downstream step in the manufacturing process.

Industry Presentations

Centocor Ortho Biotech *(P. Alfonso)*

Centocor Ortho Biotech practice is to validate viral clearance using worst-case conditions based on process proven acceptable ranges. Centocor Ortho Biotech has selected, based on scientific rationale, worst-case conditions for Protein A as minimum wash volumes (wash to baseline (WTB): 2 column volumes, intermediate wash (IW): 4.5 column volumes) and maximum width eluate collection criteria (start/stop: 50 mAU/mm pathlength). Because Centocor Ortho Biotech hasn't established what constitutes worst-case for load ratios and flow rates, the following high and low ranges representing manufacturing conditions are studied: load ratio, 50 and 5 g/L resin; flow rate, 500 and 150 cm/hr. Centocor Ortho Biotech routinely uses four viruses for spike/removal studies: Ecotropic retrovirus (ERV), PRV, Reo-3 and Polio (Table 1). Clearance was measured using infectivity assays and did not differentiate between virus removal and inactivation due to low pH during the elution.

Table 1: Clearance (infectivity) of Endogenous Retrovirus (ERV), Pseudorabies virus (PRV), Reovirus (Reo), and Poliovirus (Polio) by Protein A chromatography using proprietary mAbs "A" and "B". Data courtesy of Pedro Alfonso, Centocor Ortho Biotech.

Product	Load ratio	Flow-rate	ERV	PRV	Reo	Polio
A	High	High	>7.2	>6.4	3.3	4.1
	High	Low	7.8	4.9	4.0	4.7
	Low	High	>6.7	>6.1	2.2	4.0
	Low	Low	>7.2	>5.6	2.8	3.8
B	High	High	> 7.4	> 6.6	0.8	3.3
	High	Low	> 6.9	> 7.5	1.1	4.2
	Low	High	> 6.3	> 6.0	0.4	3.5
	Low	Low	> 6.1	> 6.3	1.0	3.3

Based on an experience-base of eight products, Centocor Ortho Biotech believes that ERV and PRV clearance and inactivation on Protein A is consistently effective, with generally >5 $\log_{10}$ LRV. Both of these are enveloped viruses, so the acid pH elution buffer undoubtedly contributes a fair level of inactivation to the overall clearance value (see discussion above). Polio virus, a small non-enveloped virus, is cleared at intermediate levels (3.3 to 4.7 $\log_{10}$), while Reo-3 virus experienced the most variable clearance, with apparent product-specific effects. For example, 2.8 to 4.0 $\log_{10}$ was cleared by the product A process; while for the product B process, only 0.4-1.1 $\log_{10}$ was cleared. It was not obvious from Centocor Ortho Biotech's data what constitutes worst-case in terms of load ratio and flow rate, supporting their practice of bracketing these parameters.

Retrovirus and poliovirus clearance after various process changes were also evaluated by Centocor Ortho Biotech (Table 2). As at most firms, Centocor Ortho Biotech implements process changes as products are refined and scaled up during development. These process changes have included cell line, cell culture media changes as well as some chromatography optimizations (listed in Table 2). Despite the diversity of changes, Centocor Ortho Biotech has found that viral clearance/inactivation remains largely consistent for Protein A unit operations as products move down the pipeline.

Table 2: Clearance (infectivity) of murine retrovirus and Poliovirus by Protein A chromatography using six proprietary mAbs. Data courtesy of Pedro Alfonso, Centocor Ortho Biotech.

Phase	Murine retrovirus	Poliovirus
	Product C	
I	4.8	3.8
II	5.7	3.4
III	>4.9, >5.2	N/A
	Product D	
I	>6.4	3.3
III	>5.9	4.3
	Product E	
I	>3.9	N/A
II	>6.1	2.4
	Product F	
I	4.5, 5.0	2.9
	Product G	
I	>6.2, >6.4	N/A
	Product H	
I	>5.0	N/A
III	>5.7	3.8

Product C: Phase I and II process changes. Phase III, with and without intermediate wash
Product D: Process changes between phases
Product E: Cell culture media change between Phases I and II
Products F and G: Lower and upper load limits evaluated
Product H: Cell line and process changes between phases

MacroGenics *(D. Farb, Ph.D.)*

MacroGenics presented comparative results from two IgG_1 humanized monoclonal antibodies produced in typical CHO expression bioreactor cultures. The Protein A step demonstrated >4 log_{10} clearance of the MuLV virus in scaled-down models, but clear loading and wash buffer conductivity effects were evident. The MacroGenics platform process uses GE Healthcare MabSelect as a protein-A capture step. To validate clearance by this step, two model viruses, A-MuLV and Reo-3, were used in spike/removal studies at a contract testing facility. Quantification of virus was by Q-PCR (MuLV) and plaque assay (Reo-3).

Previously observed subtle differences in the affinity chromatography patterns between the two mAbs provided a good opportunity for MacroGenics to explore whether a 1M NaCl wash step optimization, the monoclonal specificity, or column loading affected the clearance of two model viruses. Clearance results for the two model viruses differed, where Q-PCR measurements of MuLV removal ranged between 3-6 log_{10} and Reo-3 infectivity was reduced ~2 log_{10} (Table 3). From the limited set of data, it is not possible to identify any dependence on pH or protein titer, but the addition of a limited volume 1 M NaCl wash clearly enhanced the removal of MuLV. Similar results, >6.2 and 5.7 log_{10} MuLV clearance, were obtained for both mAbs when loaded at <45% of the binding capacity with this wash. However, when the column was loaded to a more typical 90% capacity (see Table 3 mAb 1), this apparent robustness of Protein A affinity capture clearance was compromised.

Table 3: Clearance of X-MuLV (Q-PCR) and Reo-3 (infectivity) by Protein A chromatography using proprietary mAbs "1" and "2". Data courtesy of David Farb, MacroGenics.

Process load	**Virus spike study conditions**					**LRV**	
	% Capacity	Wash	g/L load	pH load	mS/cm load	MuLV	Reo-3
mAb2 (pI~8.6)	30 30	Low salt 1M NaCl	0.57 0.57	6.4 7.3	12.5 12.1	3.14 > 6.23	ND ND
mAb1 (pI~9.1) Typical condition →	45 90	1M NaCl 1M NaCl	1.1 1.1	7.2 7.2	14.7 14.7	5.68 3.67	2.07 2.43

ND = not determined

In summary, MacroGenics found that inclusion of >1 column volume of 1 M NaCl wash in the purification platform results in a clearance factor of >3.6 $\log_{10}$ MuLV removal for a wide range of loading conditions. Underloading the column can further increase MuLV removal. The results of this clearance study point to the importance of optimizing the column wash conditions at maximum protein load when defining an effective purification platform that can be generically applied to multiple antibodies.

ImClone Systems *(M. Barry)*

ImClone presented data from four separate antibody studies (Table 4). ImClone views Protein A chromatography as a unit operation that is capable of measurable virus removal, albeit usually at modest levels. Since X-MuLV is sensitive to low pH, ImClone uses Q-PCR testing to measure X-MuLV removal by physical partitioning over Protein A, thus eliminating effects of the low pH elution step. MMV is not predicted to be affected by the low pH elution conditions (i.e. it is low pH-insensitive) and therefore the standard infectivity ($TCID_{50}$) assay can be used to assess clearance. ImClone uses MabSelect SuRe™ Protein A (GE Healthcare) for mAb capture. For validation purposes, presumed worst-case conditions are employed. This includes the maximum column height utilized in manufacturing, affording a longer residence time, potentially allowing non-specific binding of virus. Worst-case also includes product loading at the standard manufacturing limit (~70% of binding capacity), a value significantly below the total binding capacity and predicted to allow non-specific binding of potential virus in addition to specific affinity binding of the antibody to the resin. In two of the four studies, ImClone set the linear flow rate purposely lower than the manufacturing target to test viral clearance at an increased residence time (potentially increasing non-specific binding of virus).

Table 4: Clearance (Q-PCR) of X-MuLV and MMV by Protein A chromatography using four proprietary mAbs. Data courtesy of Michael Barry, ImClone Systems.

Antibody	**Flow velocity (cm/hr)**	**LRV**	
		X-MuLV	MMV
A	380 (Target)	3.53	3.8
B	380 (Target)	2.69	2.65
C	285 (Target – 25%)	2.57	2.46
D	285 (Target – 25%)	2.15	2.12

Largely comparable clearance was observed for both X-MuLV and MMV within each of four separate Protein A virus challenge studies, although slightly higher clearance levels were achieved in the two studies using a higher target flow velocity (Table 4). It could be conjectured that the lower flow velocity during the load and wash steps, associated with a longer residence time condition, could allow for higher levels of non-specific binding of virus. However, with a limited data set and the possibility that the differences in viral clearance levels could be related to the specific antibody or standard variation of data, it is not possible to reliably determine the cause for these differences. Regardless, the data demonstrate that while not as robust as other unit operations, Protein A chromatography is capable of achieving a measure of orthogonal virus clearance that may be additive to an overall process claim.

Human Genome Science *(O. Galperina)*

In HGS' experience, affinity steps are effective in clearing retroviruses (X-MuLV) routinely affording above 3 $\log_{10}$ LRV (Fig. 3). Unfortunately, HGS has not been able to match this clearance value with two other model viruses that they routinely evaluate (MMV and Reo-3; Fig. 4). The Reo-3 model, in particular, performed poorly under the process conditions investigated. An analysis of viral distribution in various column fractions was performed to pinpoint the source of the poor clearance. The mechanistic analysis suggests that X-MuLV, and to a lesser degree MMV, do not non-specifically bind to the affinity column resins. In contrast, Reo-3 does not flow through as readily during the loading phase (Fig. 5). This non-specific interaction may mechanistically explain differences in clearance results achieved with these three model viruses.

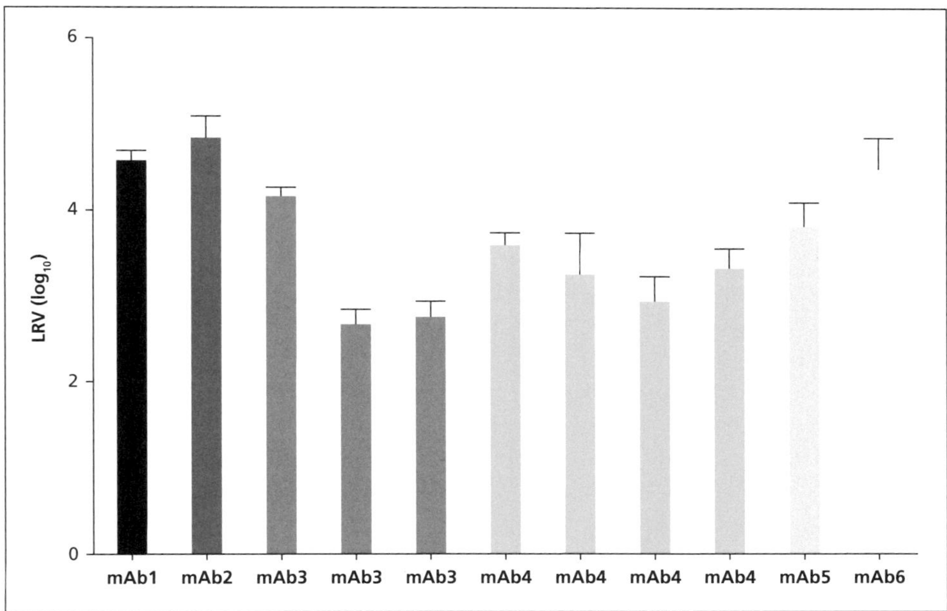

Fig. 3: Clearance of X-MuLV by Protein A chromatography runs using six proprietary mAbs (measured by Q-PCR). Data courtesy of Olga Galperina, Human Genome Sciences (HGS).

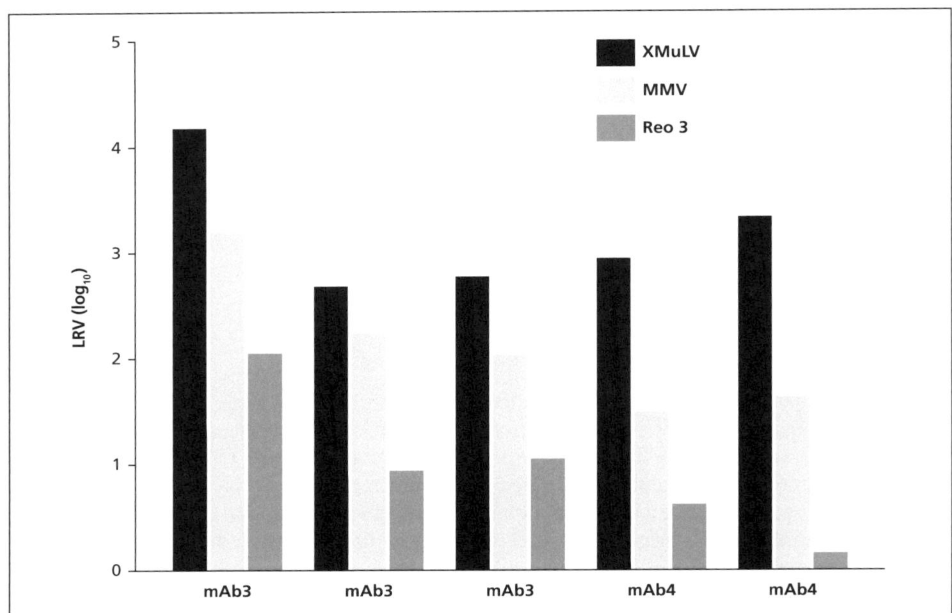

Fig. 4: Clearance of X-MuLV, MMV, and Reo-3 by Protein A chromatography runs using proprietary mAbs "3" and "4" (measured by Q-PCR). Data courtesy of Olga Galperina, Human Genome Sciences (HGS).

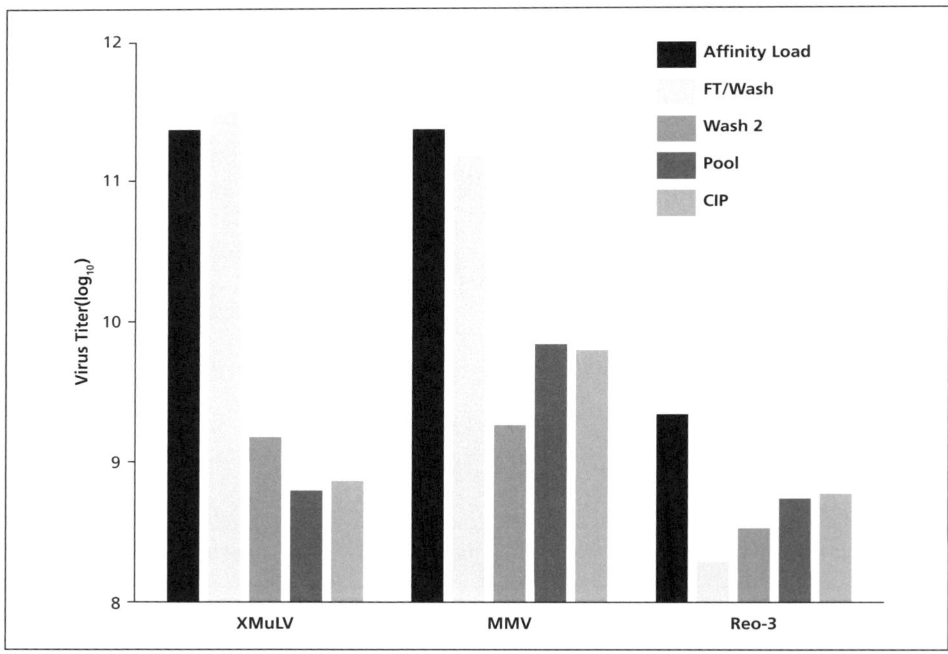

Fig. 5: Analysis of X-MuLV, MMV, and Reo-3 partitioning during affinity chromatography of undisclosed proprietary mAbs (measured by Q-PCR). Data courtesy of Olga Galperina, Human Genome Sciences (HGS).

Amgen *(B. Hubbard and L. Connell Crowley)*

Amgen has an extensive experience-base (>30 mAb processes) which shows that Protein A chromatography (MabSelect and MabSelect SuRe) can provide some level of viral clearance. Amgen has found that Protein A chromatography run under center-point operating parameters achieves 1-4 $\log_{10}$ LRV for X-MuLV (Q-PCR) and 0-3 $\log_{10}$ LRV for MMV (infectivity; Fig. 6). Although viral removal across different mAb products shows variability in log reduction of virus, it is noted that the particular clearance achieved for a process step is reproducible when replicate evaluations are performed. This argues that the observed variability is not random, rather it reflects actual product/process effects.

A cursory look at this data did not identify any correlation between viral removal and any particular process parameter. Process parameters where no clear correlation with virus removal could be found include the mAb product in the process fluid itself, the column wash volume, and the elution pH. For example, Figure 7 shows the removal of X-MuLV on Protein A chromatography for six different mAb products run under identical process conditions with the exception of the chromatography resin (MabSelect or MabSelect Sure resin) and the elution pH (ranging from 3.4 to 3.7). In each case, viral removal was similar (PCR measured LRV 2.0-2.5), showing no evidence for a mAb product or elution pH effect.

An analysis of X-MuLV removal obtained as a function of the Protein A column wash pH and composition is shown in Figure 8. While no dramatic improvements in X-MuLV removal were seen, some column wash regimens seem to improve viral clearance modestly (e.g. arginine or propylglycol-based washes). This argues that a comprehensive evaluation of wash components, wash salt strength, and column loading may be a promising avenue to improve X-MuLV clearance.

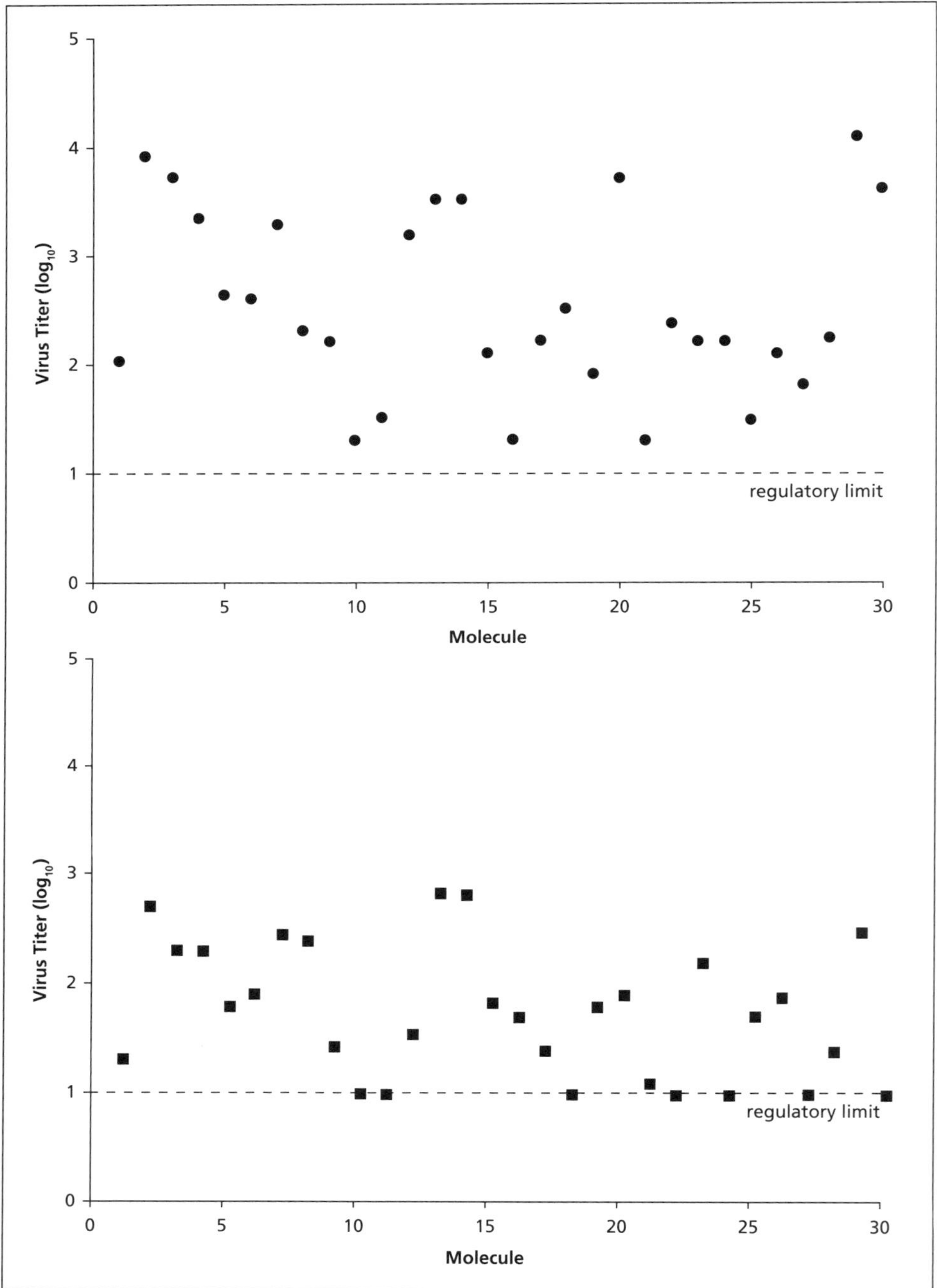

Fig. 6: Clearance of X-MuLV (circles) and MMV (squares) by Protein A chromatography using undisclosed proprietary mAbs. Data courtesy of Brian Hubbard and Lisa Connell Crowley, Amgen.

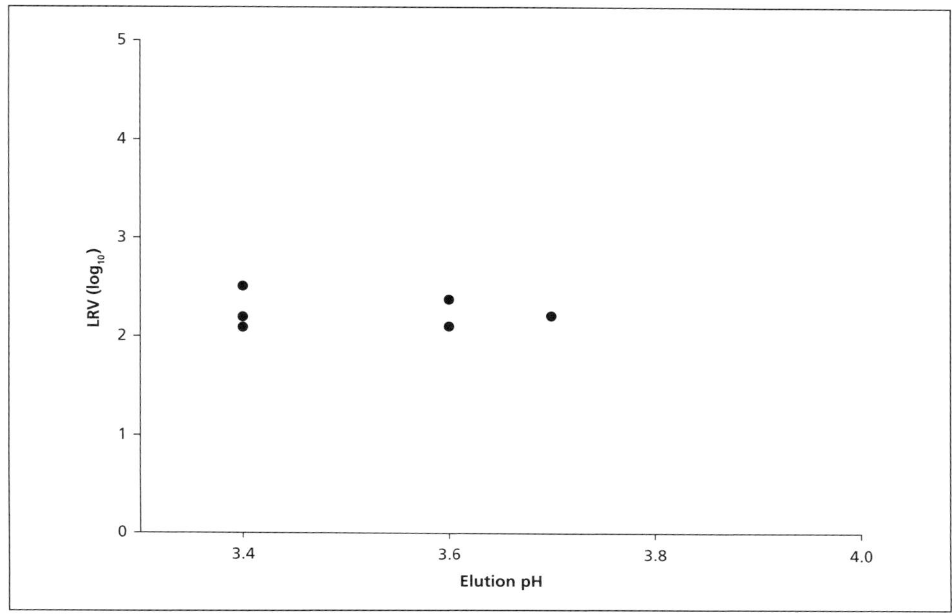

Fig. 7: Clearance of X-MuLV by Protein A chromatography using six different proprietary mAbs run under identical operating conditions, except chromatography resin and elution pH. See text for details. Data courtesy of Brian Hubbard and Lisa Connell Crowley, Amgen.

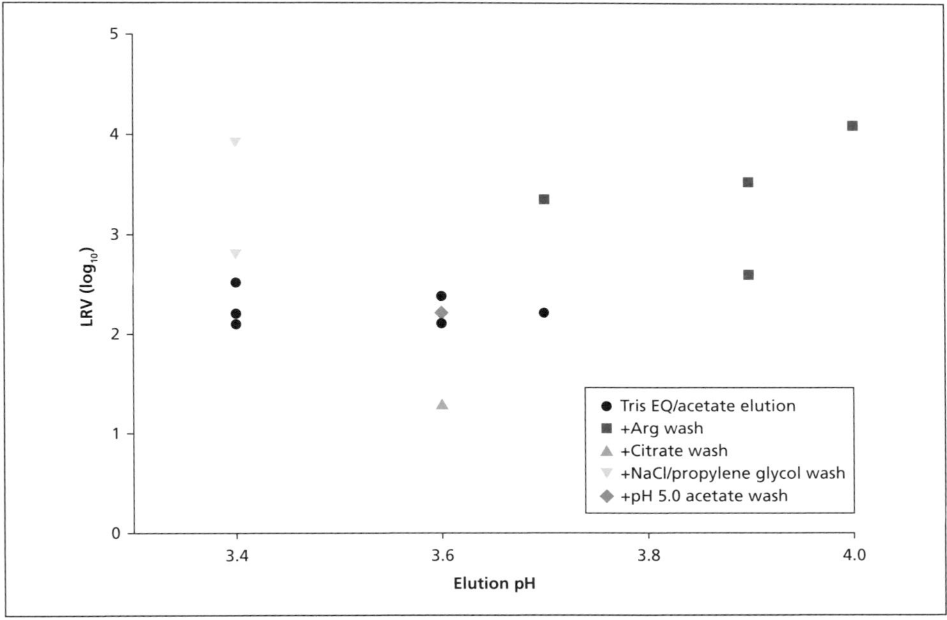

Fig. 8: Clearance of X-MuLV by Protein A chromatography using undisclosed proprietary mAbs under different column wash (i.e. buffer type) and elution (pH) conditions. Data courtesy of Brian Hubbard and Lisa Connell Crowley, Amgen.

Novartis *(M. Heitzmann)*

Novartis' virus removal validation strategy is based on previous observations and practical considerations. For phase I-III processes, virus clearance validation is usually not performed for Protein A, unless issues are encountered with validating other steps (i.e. low LRV are obtained for retrovirus). This strategy is taken because Novartis has observed only modest (2-3 log_{10}) reduction of MuLV (Q-PCR) and other viruses (Table 5). In fact, the Protein A step has a level of virus removal efficiency roughly similar to HCP, (i.e. 2 to 4 log_{10}), similar to that observed by others [7]. One outlier study was a reuse study for mAb2; in this study a repeat run with naive resin obtained a >5 log_{10} LRV. While this result was analytically valid, it was difficult to explain from a conceptual basis. The re-use part of the same study, as well as a highly similar precedent study with the same mAb, showed the "normal" 2-3 log_{10}. Based on this experience, the Protein A step is validated in most cases for Novartis launch processes only; but DoE or design space studies are usually not performed or planned (a higher priority would be placed on efficient steps like low pH, AEX and virus filtration).

Table 5: Clearance (Q-PCR) of X-MuLV by Protein A chromatography using eight proprietary mAbs. Data courtesy of Markus Heitzmann, Novartis.

Antibody	**X-MuLV**
mAb A	2.2
mAb B	2.3
mAb 1	2.6
mAb 2[a]	3
mAb 2[b]	>5.2

[a] Original study

[b] Repeat study

Pfizer *(P. Mensah)*

Pfizer presented experience with seven mAbs. Pfizer has developed a mAb platform process using MabSelect Protein A. They routinely validate their processes using X-MuLV (Q-PCR) and MMV (infectivity). Over their process development history, they have contracted studies to two different contract testing organizations (CTOs). Another key difference between their processes with a potential to impact viral clearance is elution pH, varying from 3.5 to 3.7. Based on this experience-base, Pfizer wanted to address if differences in LRV for the various mAbs could be attributed to factors such as CTO or elution pH. The data provides suggestive evidence that X-MuLV LRV may be generally higher at CTO 1 than at CTO 2 (Table 6). Direct comparison studies will be needed to confirm the reality of that hypothesis, but Pfizer believes that the subtle trend may due to the level of sensitivity of Q-PCR assay or other CTO-specific factors like the virus preparations.

Table 6: Clearance (Q-PCR) of X-MuLV and MMV (infectivity) by Protein A chromatography using seven proprietary mAbs. Data courtesy of Paul Mensah, Pfizer.

mAb	**CTO**	**Elution pH**	**LRV**	
			X-MuLV	MMV
A	1	3.7	3.48	2.84
B	1	3.5	3.48	1.66
E	1	3.5	5.09	2.8
C	2	3.7	2.24	1.9
D	2	3.7	1.4	1.33
F	2	3.7	2.94	1.81
G	2	3.5	2.5	1.03

A larger analysis was also performed on historical data from fourteen MabSelect processes and earlier processes using other resins. Pfizer found no evidence of a significant difference in LRV between the 3 Protein A resins (Table 7).

Table 7: Clearance (Q-PCR) of X-MuLV and MMV (infectivity) by three different Protein A chromatography resins. Data courtesy of Paul Mensah, Pfizer.

Protein A Type	# of mAbs	X-MuLV	MMV
rmp Protein A	4	3.85 ± 1.22	3.78 ± 0.48
Protein A FF	3	3.69 ± 1.21	2.91 ± 1.05
MabSelect	14	3.11 ± 0.72	2.10 ± 0.64

Regeneron: Experience managing CTO relations *(J. Mattila)*

Regeneron presented a useful case study for managing contract testing organization (CTO) relations. This case involved an Fc fusion protein product with a variable Protein A clearance capability; further investigation (described below) ultimately led to strategies to avoid potentially anomalous results in the future. Regeneron performs viral clearance

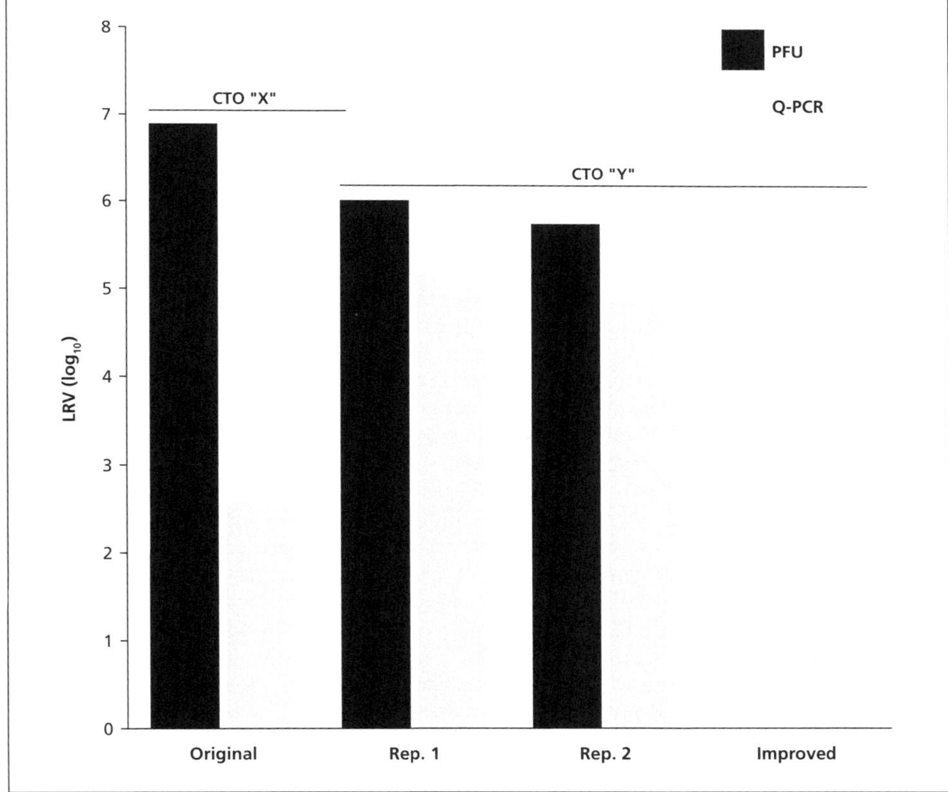

Fig. 9: Improvement in measured clearance (both infectivity and Q-PCR) of X-MuLV by Protein A chromatography after implementing an improved extraction procedure at a second CTO. Data courtesy of John Mattila, Regeneron.

studies supporting 2+ INDs per year and has experience with multiple CTOs. In this case, they obtained expected X-MuLV clearance by infectivity and Q-PCR at CTO "X" (Fig. 9). However, in repeat testing, much greater clearance (as measured by Q-PCR) was obtained at CTO "Y". As a follow-up, repeat testing was performed at CTO "Y", replicating the anomalous result.

Their investigation focused on CTO "Y" practices, particularly the Q-PCR assay limit of detection (known through interference testing) and differences in multiple validated assays used by CTO "Y". They identified a potential root cause for the anomalous results and discussed changes to the extraction method resulting in improved viral RNA recovery in product pool. A confirmatory study yielded the more expected result, previously seen from CTO "X" (Fig. 9). The lesson of this exercise was that close cooperation with contract virus labs is required, and it is critical to understand factors affecting assay outcome. In summary, exerting ownership of your (firm's) process is a key to success.

Overall Summary, Protein A

The session concluded that Protein A is an operationally complex unit operation (i.e. complex feedstocks, multiple buffer phases) and that this complexity extends to viral clearance. For example, many firms found that parameter effects may exist, but are hard to explain. Variation in LRV between products/processes appears to be a real phenomenon, but does not seem to exist within the same product/process. This suggests that much of the variation results from feedstock complexity, with an unknown contribution from the large overall number of phases (i.e. equilibration, load, wash, secondary wash, elution, cleaning). Given the bind and elute nature of the step, nonspecific adsorption of virus on resin plays a major role in LRV, but this is hard to model in lab studies. This raises questions pertaining to validation, for example: (1) How does one model worst-case when the exact viral clearance mechanism (and effects of certain parameters) is unknown? and (2) As a consequence, if worst case is not well understood, what is the most appropriate way to challenge the Protein A chromatography process for viral reduction (i.e. what are appropriate target conditions if this is the case)?

Some consistent trends were noted across the industry presentations, the regulatory database analysis, and from existing scientific literature. For example, the low pH inactivation component of Protein A chromatography is important for clearance of enveloped viruses. Symposium participants consistently found that some viruses partition better than others (e.g. X-MuLV > MMV > Reo-3). In fact, Reovirus was generally found to be a problematic virus for clearance, perhaps because of preparation quality issues (preparations tend to be heterogeneous and difficult to purify). Viral clearance does not decrease after extensive re-use (industry experience and published data), again probably because of the BE nature of Protein A chromatography. Finally, when performing validation studies at a contract site, it is critical to carefully monitor and exert ownership over the product/process.

Key areas for process improvement and directed investigation identified at the symposium focused on clarifying how viruses are cleared by Protein A. For example, investigating mechanisms of product vs. resin binding of virus or feedstock effects (e.g. blocking non-specific binding) may identify what contributes to virus "stickiness" and how to reduce this phenomenon and improve LRV. This whole area of investigation, however, could be leapfrogged if strategies for tracking retrovirus-like particles directly at large scale are developed [8]. If this can be accomplished economically, for example with PCR-based methods, it would likely constitute an improvement over spike/removal validation.

Miesegaes G, Bailey M, Willkommen H, Chen Q, Roush D, Blümel J, Brorson K (eds): Proceedings of the 2009 Viral Clearance Symposium. Dev Biol (Basel). Basel, Karger, 2010, vol 133, p 23.

SESSION III

Low pH Treatment

Miesegaes G, Bailey M, Willkommen H, Chen Q, Roush D, Blümel J, Brorson K (eds): Proceedings of the 2009 Viral Clearance Symposium. Dev Biol (Basel). Basel, Karger, 2010, vol 133, pp 25-42.

Low pH Treatment

Technical Background

Exposure of process streams to low pH (i.e. pH levels < 4) has been used as a method for retrovirus and other enveloped virus inactivation for more than 20 years. In a mAb purification process, the capture Protein A affinity chromatography step often employs a low pH elution buffer. Low pH hold in the Protein A pool therefore is widely used as a dedicated viral inactivation step. It has been shown that the loss of retroviral infectivity by low pH is due to changes of virus capsid and lipid envelope structure with no impact to virus reverse transcriptase activity and viral genomic RNA [9].The pH, temperature, and time used in a typical Protein A pool for inactivation of retrovirus and other enveloped viruses is not expected to be effective to inactivate small, non-enveloped viruses such as MMV [10].

Regulatory Agency Presentations

FDA *(G. Miesegaes)*

Chemical inactivation techniques presumably damage viral envelopes or envelope proteins. Therefore, non-enveloped viruses are not expected to be dramatically impacted. This notion was supported through a comparison of LRV and virus type; whereas MuLV and PRV were effectively inactivated, parvo-and picornaviruses (e.g. Polio) for instance were not [3]. Because of this, we focused our analysis exclusively on MuLV (an enveloped retrovirus).

It was found that low pH inactivation was quite effective in clearing MuLV, although claims of complete clearance (as defined by a ">x $\log_{10}$" value reported in the regulatory submission) were not made in all cases (incomplete: n=105; complete: n=149; Fig. 10a-b). The majority of records reported successful inactivation (5-8 $\log_{10}$) of MuLV; only a small number (~3%) of records reported LRV <2 $\log_{10}$. These latter reports were defined as "outliers" and upon further review were found to correlate with the use of a higher pH [3]. Specifically, records containing <2 $\log_{10}$ LRV on average had employed a pH of 3.9, whereas records containing >6 $\log_{10}$ LRV employed an average pH of 3.4.

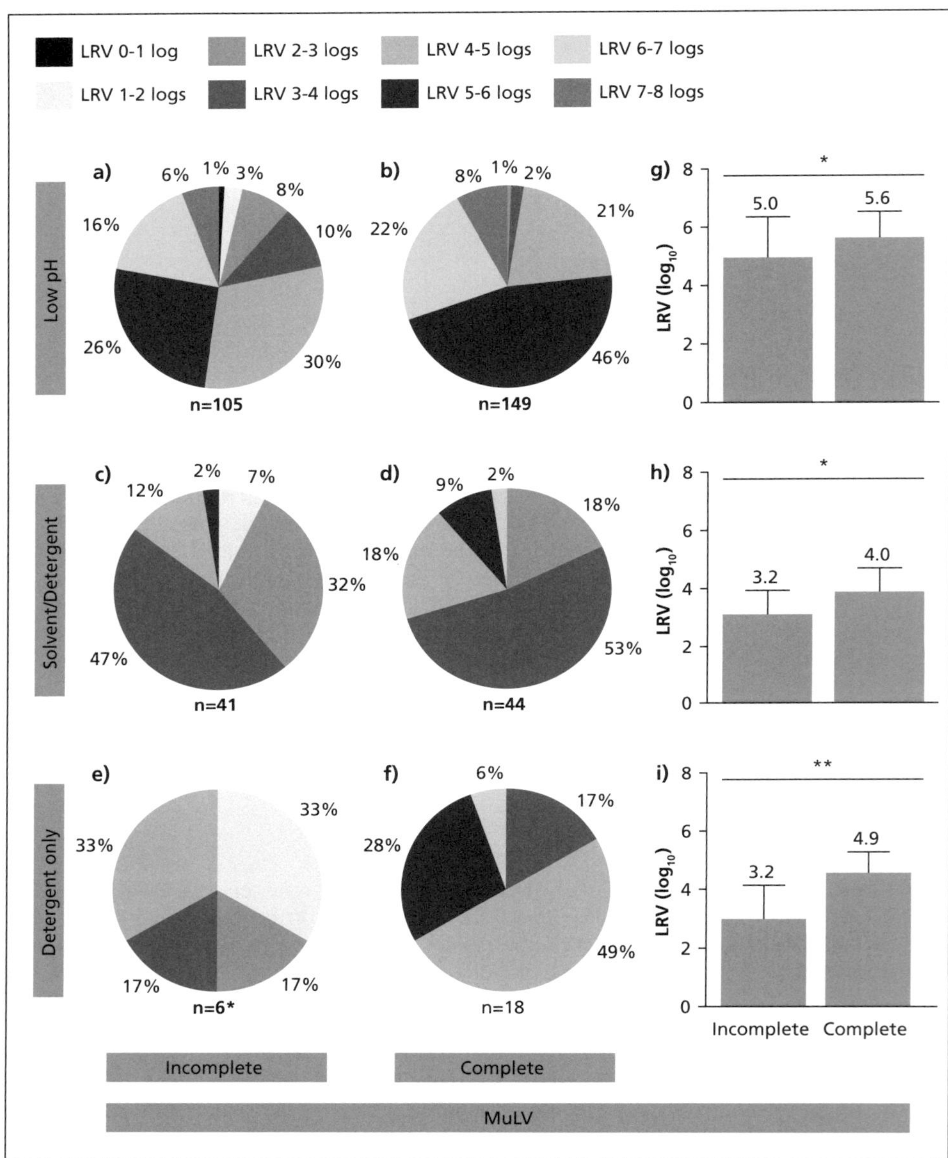

Fig. 10: Analysis of inactivation step records in CDER's viral clearance database. (a-b) low pH, (c-d) solvent/detergent, and (e-f) detergent-only inactivation steps. Pie charts show MuLV records binned into 1 log_{10} LRV ranges. For this analysis, records where claims of complete clearance ("complete"; b, d, f) were separated from those where such a claim was not made ("incomplete"; a, c, e). Mean LRV (g-i) for each are shown. Bars are standard deviations. *p<0.0001; **p=0.0007. Data courtesy of George Miesegaes, FDA.

PEI *(J. Blümel)*

Inactivation of retroviruses by incubation of the production intermediate at low pH (pH 3.5 to 4.0) was investigated by individual companies, which included 72 studies from 91 different antibodies. The low pH intermediates represented the eluate from Protein A (or similar) chromatography applied to the clarified cell culture harvest fluids. Retroviruses were spiked into these intermediates. Most investigations (63 cases) were performed using X-MuLV as a model retrovirus. A-MuLV (4 cases) or ecotrophic E-MuLV (4 cases) were occasionally used. In 63 cases, retrovirus inactivation factors above 4.0 $\log_{10}$ were determined. In these cases, inactivation was mostly below the detection limit while residual infectivity was occasionally detected in 14 from these 63 cases. Nine of 72 antibodies (12.5%) were identified where logarithmic retrovirus inactivation was limited below a value of 4.0 and residual infectivity was detected. Analysis of these nine cases is summarized in Table 8. All studies were performed by the same contract laboratory organization which was also involved in producing most of the other data showing efficient retrovirus inactivation. Two of these study reports clarified that the spiking virus had been purified by ultracentrifugation. While such information was lacking in other reports it might be assumed that ultra-centrifuged X-MuLV was also used in the other cases because all studies were from the same organization. In some cases, no obvious difference from the conditions described for effective retrovirus inactivation, i.e. pH ≤3.8; incubation time ≥30 min, acetate or citrate buffer, pI between aprox. 3 and 9; NaCl ≤500mM, protein concentration ≤40 mg/mL. [9], could be identified. However, there might be other relevant factors which were not outlined in the validation study reports.

Table 8: Viral clearance studies by low pH treatment where inactivation of retroviruses (X-MuLV) was limited to below 4.0 $\log_{10}$. Data courtesy of Johannes Blümel, PEI.

Case No.	LRV	pH at validation study	Time [min]	Temp. [°C]	Production cells	Buffer-System	mAb pI
1	2.99	3.7±0.1	60	20 ± 5	CHO	50mM sodium sulfate 100mM acetate/ acetic acid	8.9
2	1.9	3.75-3.80	60	20	mouse hybridoma	10mM sodium formate/ acetic acid	5.9 to 7.4
3	2.97	3.8	60	22 ± 1	CHO	50mM sodium acetate/ acetic acid	5.8
4	3.91	3.79	60	RT[a]	NS0	100mM sodium citrate/ citric acid	7.2 to 7.9
5	3.54	3.7	120	2-8	CHO	25mM glycine 3mM phosphoric acid/ formic acid	8.2
6	3.56	3.75	60	19 ± 4	CHO	10mM sodium formate/ acetic acid	NI[b]
7	2.1	3.76	60	RT[a]	NS0	acetic acid	9
8	2.62	3.8	60	RT[a]	CHO	100mM acetic acid/ acetic acid	9.5
9	1.00	3.8	60	21 ± 3	NS0	sodium citrate/ citric acid	8.2

[a] RT: room temperature

[b] NI: no information

In summary, inactivation of X-MuLV seems robust at pH <3.8, >15°C, for 30-60 min. Nevertheless, the exact conditions for virus inactivation need to be carefully evaluated and it might be prudent to perform confirmatory experiments on virus inactivation whenever the relevance of previously-obtained data is not totally clear.

Industry Presentations

Amgen *(B. Hubbard and L. Connell Crowley)*

Low pH inactivation is commonly used as a dedicated viral safety step in monoclonal antibody (mAb) manufacturing. Data from >30 Amgen mAb processes indicate that the low pH inactivation step is highly robust for inactivation of enveloped viruses with the majority of studies achieving complete inactivation of X-MuLV at pH $\leq$ 3.8 (Fig. 11).

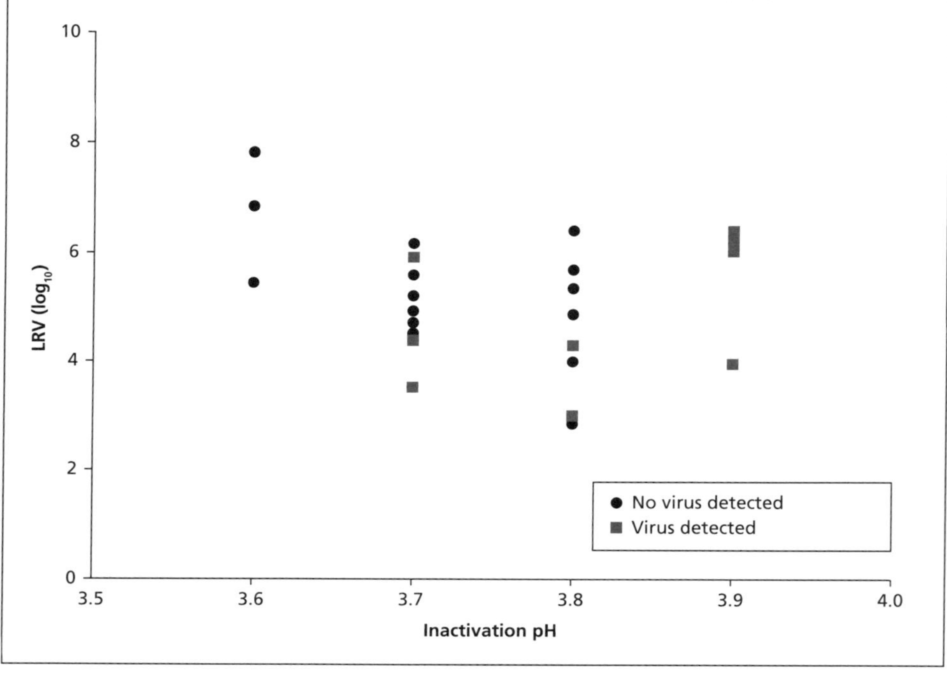

Fig. 11: Effect of pH on X-MuLV inactivation by low pH treatment using undisclosed proprietary mAbs. Acidification solutions used included acetic, phosphoric, citric, hydrochloric and formic acids. The buffering species included acetate, citrate and glycine at protein concentrations between ~4-30 g/L. The duration of low pH inactivation ranged from 30-120 minutes. The majority of studies achieve complete clearance at pH $\leq$ 3.8. Data courtesy of Brian Hubbard, Tom McNerney and Lisa Connell Crowley, Amgen.

An analysis of Amgen's database was performed to determine whether the acidification solution, buffer type, mAb product or mAb protein concentration affected the LRV obtained for X-MuLV (pH 3.7, 60 minutes). No correlation of X-MuLV clearance was found for buffer type or concentration when 25 – 100mM acetate or 50 – 100mM citrate was used (Table 9). In addition, the X-MuLV LRV did not correlate to the type of acidification solution when acetic, phosphoric or formic acids were used or to the mAb type or mAb concentration between ~4-30 g/L.

Table 9: Clearance of X-MuLV by low pH (3.7, 60 min incubation time) treatment using 19 undisclosed proprietary mAbs. Data courtesy of Brian Hubbard and Lisa Connell Crowley, Amgen.

Molecule	Buffer species	Buffer molarity	Acid used	X-MuLV
1	acetate	25 mM	phosphoric	5.14
2		50 mM	acetic	>5.61
3				>6.24
4			formic	>6.49
5				>6.92
6		100 mM	acetic	>4.51
7				>5.97
8				>6.17
9				>5.74
10				>4.72
11				>5.97
12			phosphoric	>5.56
13				>5.14
14				>5.14
15				>6.12
16				4.4
17	citrate	50 mM	citric	>5.18
18		100 mM	phosphoric	>4.92
19	glycine	25 mM	formic	3.5

Four of the clearance studies did not achieve the expected clearance of >4.6 Log_{10} that resulted under conditions of ≤ pH 3.8 shown in the bracketed studies reported by other firms [9]. Amgen has performed an analysis of three of the outliers. Repeat runs of two of the outliers achieved complete inactivation of the virus with LRVs >5 logs, suggesting that technical issues may be responsible for the unexpectedly low clearances values obtained during the initial studies.

The data from a third outlier was repeatable, and further study indicated that glycine, which was used as the buffering species, can slow the inactivation kinetics of X-MuLV at pH 3.7. This effect appears to be concentration dependent, with X-MuLV inactivation kinetics slowing as glycine concentrations are decreased (Fig. 12). This buffer-specific kinetic effect appears independent of model protein type or concentration in that no model protein was present in the experiment. The effect of glycine on the kinetics of inactivation can be overcome by reducing the pH of the inactivation to pH 3.6 (Fig. 13).

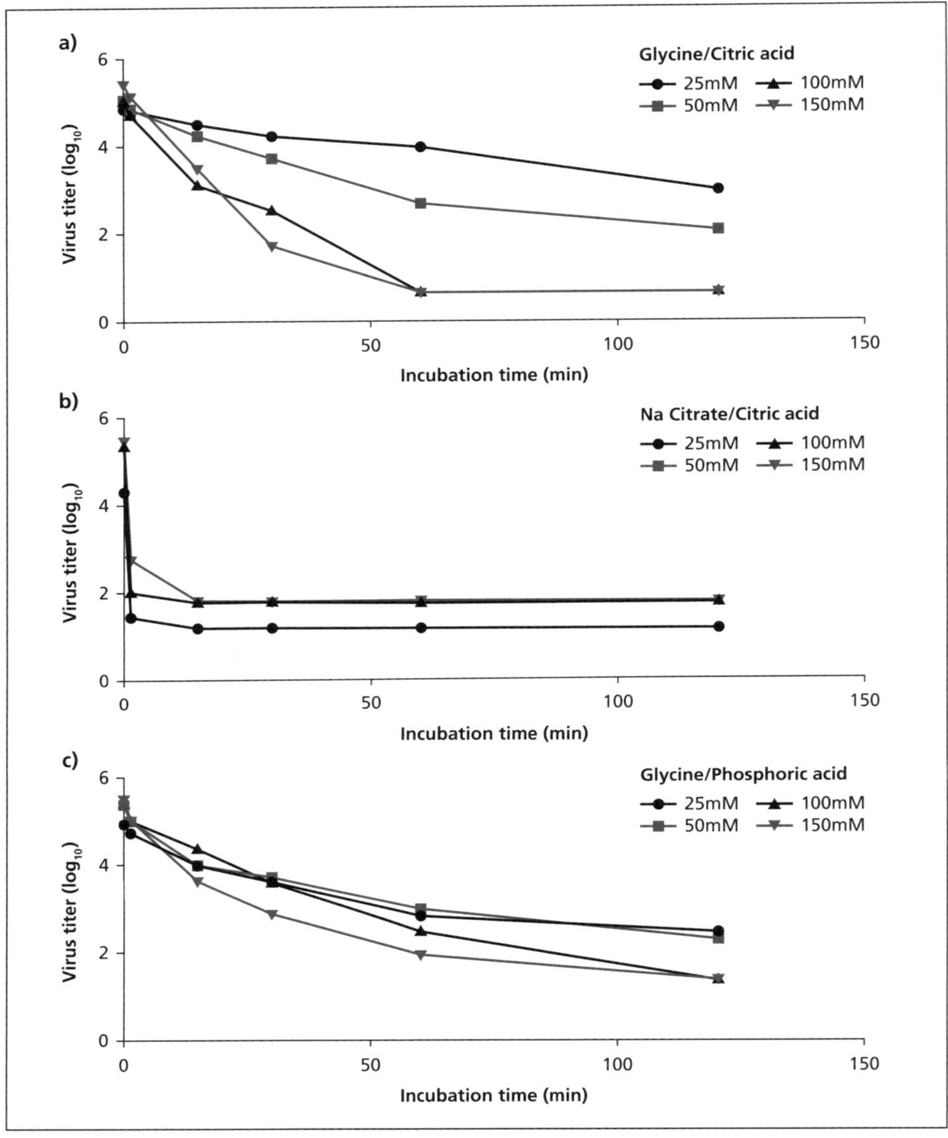

Fig. 12: Effect of buffer on X-MuLV inactivation by low pH (3.7) treatment in the absence of model protein under a range of buffer conditions. In the presence of glycine (panels A and C), X-MuLV inactivation appears to slow as the glycine concentration is decreased. Similar slow kinetics were not seen with citrate buffers (panel B). Data courtesy of Brian Hubbard, Tom McNerney and Lisa Connell Crowley, Amgen.

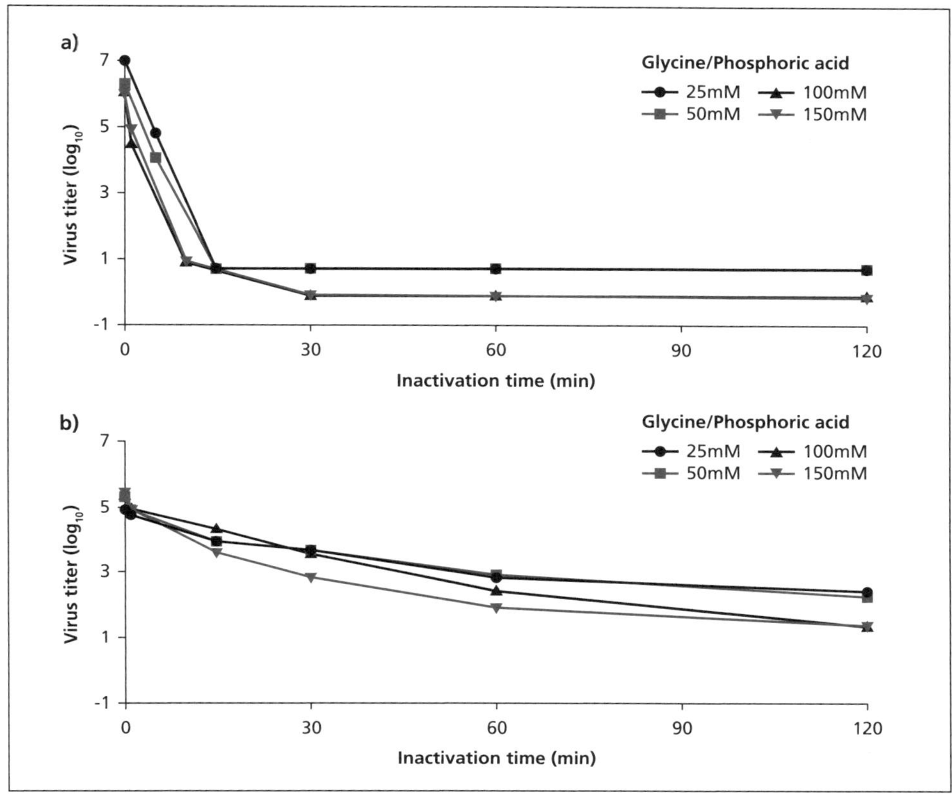

Fig. 13: Clearance of X-MuLV by low pH treatment in glycine buffer acidified with phosphoric acid. Half of the experiments were performed with and half were performed without mAb. The slow inactivation kinetics of glycine at pH 3.7 (panel B) can be overcome by reducing the pH to 3.6 (panel A). Data courtesy of Brian Hubbard, Tom McNerney and Lisa Connell Crowley, Amgen.

Abbott Bioresearch Center *(E.O. Lundell)*

The low pH inactivation of the enveloped virus X-MuLV has been evaluated under differing processing conditions on three monoclonal antibodies with pIs ranging from 8.8 to 9.3. Inactivation was evaluated in a sodium acetate/phosphate matrix. The antibody concentration varied from 15 g/L to 25 g/L, and the incubation pH varied from 3.5 ± 0.1 to 4.0 ± 0.1. All inactivation studies were performed at ambient temperature (18 - 23°C); results are given in Table 10. Significant inactivation occurred at pH 3.5 ± 0.1 by the time the first sample was taken (within minutes of pH adjustment), and all inactivations performed at pH 3.5 ± 0.1 were complete to the limit of detection after 15 minutes under normal sampling conditions. Thus, a claim for 60 minutes at this pH would have a built-in 45 minute buffer time, speaking to the robustness of the step.

Table 10: Clearance of X-MuLV by low pH treatment using three proprietary mAbs. Data courtesy of Edwin O. Lundell, Abbott Bioresearch Center.

mAb	pI	Matrix	Acid	pH	[mAb]	Time	X-MuLV[a]
A	9.3	Na acetate	0.5 M H_3PO_4	3.5 ± 0.1	25 g/L	60 min	≥ 5.46 ± 0.07 ≥ 5.53 ± 0.10
B	8.8	Na acetate	0.5 M H_3PO_4	3.5 ± 0.1	15 g/L	60 min	≥ 5.45 ± 0.10 ≥ 5.45 ± 0.10
C	~9.0	Na acetate	0.5 M H_3PO_4	3.5 ± 0.1	21 g/L	60 min	≥ 4.90 ± 0.16 ≥ 4.90 ± 0.16
			0.5 M H_3PO_4	4.0 ± 0.1	21 g/L	60 min	≥ 4.89 ± 0.16

[a] LRV ± 95% confidence interval.

The kinetics of inactivation of X-MuLV at pH 3.5 ± 0.1 and 4.0 ± 0.1 for mAb-D are given in Figure 14. The data demonstrate a very rapid inactivation to the limit of detection within 5 minutes at pH 3.5 ± 0.1 whereas inactivation at pH 4.0 ± 0.1 reached the limit of detection after 30 minutes.

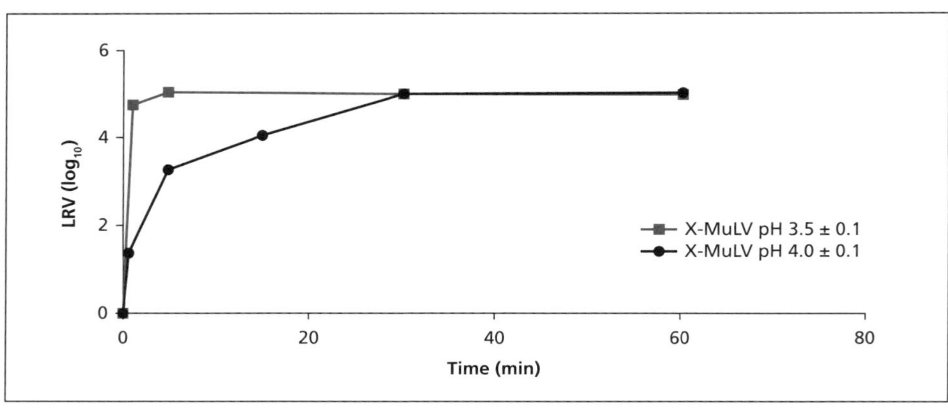

Fig. 14: Kinetics of inactivation of X-MuLV using proprietary mAb "D". Data courtesy of Edwin O. Lundell, Abbott Bioresearch Center.

Brorson et al [9] proposed a "bracketed generic clearance" LRV of $\geq 4.6 \log_{10}$ for X-MuLV based on a study of four mAbs and two other recombinant proteins. The results from the mAbs in this study strongly support the proposed bracketed generic clearance conditions for inactivation of a rodent type C retrovirus, which are pH ≤ 3.8, time ≥ 30 minutes, temperature ≥ 14°C, a buffer system of citrate or acetate (and in this study, phosphate), a protein concentration ≤ 40 mg/mL, NaCl ≤ 500 mM and a pI between approximately 3 and 9 (in this study 8.4-9.3). In addition, incubation at pH 4.0 ± 0.1 also reached the limit of detection 30 minutes, suggesting that this higher pH may also be a useful inactivation condition for antibodies that are unstable at the lower pH, particularly if the minimum time is increased to 60 minutes.

ImClone Systems *(M. Barry)*

Low pH treatment, especially as part of an antibody purification process, is widely used in the industry and considered a significant virus inactivation step. However, antibodies vary in their susceptibility to low pH conditions; some conditions can adversely affect critical product quality attributes such as protein purity including an increase in low and/or high molecular weight species (e.g. fragments and aggregates). DoE is currently being used to determine limits or worst-case conditions affecting critical product quality attributes during the low pH process including the minimum allowable pH. This assessment of pH sensitivity is performed in conjunction with other worst-case process conditions affecting product quality such as temperature, product concentration and time so that an effective design space can be defined for the low pH process. Once the minimum allowable pH for a specific antibody is determined so that acceptable product quality is maintained, an allowable pH range can then be determined to provide sufficient flexibility for manufacturing operations. This allowable manufacturing range then defines the maximum pH for the process; a pH that is worst-case with respect to the ability to inactivate virus. This maximum pH along with a combination of other worst-case parameters relating to virus inactivation can then be challenged in a single study to validate the process robustness without the need to individually challenge each parameter separately. Within our platform approach, the other worst-case factors for virus inactivation include lowest temperature (14.5 ± 0.5 °C to support a minimum of 15 °C), highest product concentration (determined by the specific antibody process) and minimum time (performed using a time course to determine inactivation kinetics up to the minimum allowable manufacturing time; 0, 5, 30 and 60 minutes).

Data representing the virus inactivation Log Reduction Value (LRV) from a variety of antibody processes using a maximum allowable low pH in the range of 3.5 – 3.9 are shown in Table 11. As would be expected, the higher the maximum allowable low pH challenged, the lower the virus inactivation. The results show that at a low pH of < 3.7, an LRV of $> 6 \log_{10}$ is achieved and usually the inactivation is complete to the limit of assay detection (LOD). At a pH of ≥ 3.7, an LRV of $> 5 \log_{10}$ is still observed, however the inactivation is usually not complete to the LOD. Additionally, slower inactivation kinetics were observed for the higher maximum allowable pH challenge studies. The kinetic data are not shown; however at a low pH of < 3.7, the limit of detection (LOD) was typically reached within 5 minutes. For a pH range of 3.7 to 3.8, an LRV of 1-3 $\log_{10}$ was typically observed by 5 minutes and the LOD or maximal clearance was reached by 30 minutes. Finally, for the studies at a pH of 3.9, an LRV of 0-1 $\log_{10}$ was achieved within 5 minutes; an LRV of approximately 4 $\log_{10}$ was achieved within 30 minutes and the LOD or maximal clearance was achieved by the final 60 minute time point.

Table 11: Clearance of X-MuLV by low pH treatment for 60 minutes using nine proprietary mAbs "A" through "I". Data courtesy of Michael Barry, ImClone Systems.

MAb	pH*	X-MuLV		MAb	pH*	X-MuLV
A	3.5	≥ 5.59		E	3.65	≥ 6.19
A	3.5	≥ 6.56		F	3.7	≥ 6.43
A	3.5	7.12		G	3.7	5.21
A	3.5	6.99		G	3.7	5.83
A	3.5	≥ 5.11		G	3.7	5.44
B	3.5	≥ 5.18		H	3.7	5.05
C	3.5	≥ 6.67		D	3.8	5.38
D	3.5	≥ 5.33		I	3.8	≥ 5.68
E	3.5	≥ 6.06		I	3.9	≥ 5.81
F	3.6	6.62		I	3.9	5.16

*pH value was highest allowable for manufacturing; therefore, challenge studies run at or above pH listed in table to validate worst-case

Overall, the results show that the specific low pH value does directly impact both inactivation kinetics and overall clearance claims as would be expected. However, the data demonstrate that low pH treatment can still be considered a very robust virus clearance process when pH < 3.7 and effective even when challenged using a combination of worst-case operating conditions. Antibodies that are more susceptible to low pH process conditions and therefore require the use of a higher pH range for the low pH inactivation process can still be effective if sufficient incubation time is built into the process.

Genentech, Inc. *(Q. Chen and L. Norling)*

In the past, Genentech has evaluated various parameters for low pH inactivation and found that within the range studied, parameters such as protein concentration, salt concentration, Protein A elution buffer, impurity level in the Protein A pool, and aggregate level do not impact X-MuLV inactivation [9].

To understand which process parameters would affect viral inactivation, we evaluated a range of pH, temperature and time using mAb 4 Protein A pools. The data show that at pH 3.6 and 3.7 and temperatures of 15-30°C, retrovirus inactivation is rapid and effective (Fig. 15a and b). At pH 3.8, inactivation occurs at a slower rate at 15 °C (Fig. 15c), whereas inactivation

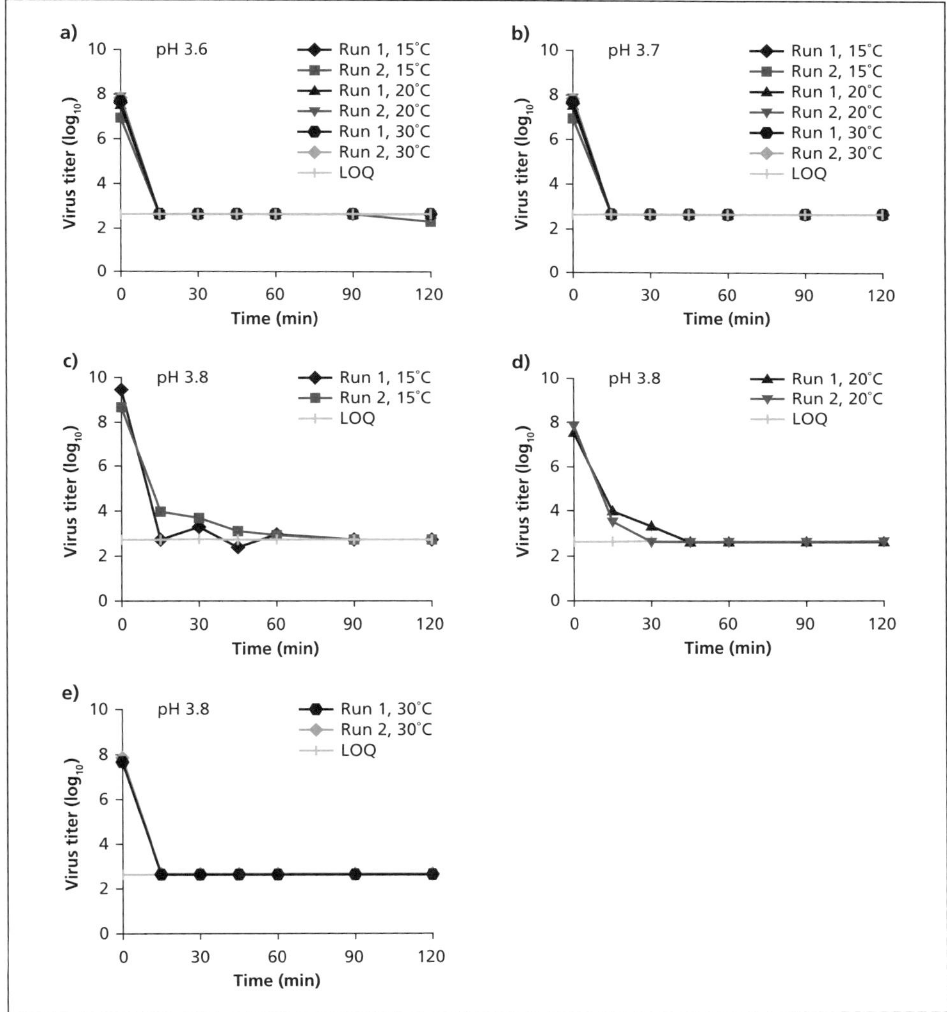

Fig. 15: Effect of pH, temperature, and time on X-MuLV inactivation by low pH using proprietary mAb "4". (a) pH 3.6; (b) pH 3.7; (c) pH 3.8 at 15°C; (d) pH 3.8 at 20°C; (e) pH 3.8 at 30°C. Data courtesy of Qi Chen and Lenore Norling, Genentech, Inc.

rate increases with increasing temperatures (Fig. 15d and e). These results support that pH, time, and temperature are parameters that can interact with one another in impacting low pH viral inactivation. In addition, lower pH, higher temperature, and longer hold time are better-case conditions. The data also suggests that pH 3.6 is most robust where effective inactivation occurs rapidly at a room temperature range of 15-30°C.

Experiments were performed to understand the impact of impurity level on low pH viral inactivation. Table 12 shows results from mAb 4 and mAb 2 Protein A pools spiked with 5% of respective HCCF where the low pH hold was at pH 3.6, 20°C for 30 minutes. The comparable inactivation, observed with and without the HCCF spike for both products, suggests that low pH inactivation under the conditions tested at the unit operation end-point time is robust with respect to impurity level in the Protein A pool.

Table 12: X-MuLV low pH inactivation is not affected by HCCF spiking using proprietary mAbs "2" and "4". Data courtesy of Qi Chen and Lenore Norling, Genentech, Inc.

Process intermediate for low pH inactivation	**X-MuLV[a]**
mAb 4 Protein A pool	>4.0
mAb 4 Protein A pool + 5% mAb 4 HCCF	>4.4
mAb 2 Phase III Protein A pool	>4.1
mAb 2 Phase III Protein A pool + 5% mAb 2 HCCF	>4.1

[a] Inactivation conditions: pH 3.6, 20°C, 30 minutes

A potential result of viral contamination in a bioreactor is an impact on cell viability. Low cell viability may lead to high membrane fragment content; this is believed to protect virus from inactivation, and thus is a potential worst-case. We therefore tested mAb4 Protein A pools purified from HCCF lots with a range of cell viability at the time of harvest (Table 13). The results show that LRVs are not affected even if using a Protein A pool purified from a HCCF lot with cell viability at harvest as low as 7%.

Table 13: X-MuLV low pH inactivation is not affected by cell viability at harvest using a proprietary mAb "4". Data courtesy of Qi Chen and Lenore Norling, Genentech, Inc.

Process intermediate for low pH inactivation	**Cell viability at harvest (%)**	**X-MuLV[a]**
mAb 4 Protein A pool lot 1	7	≥4.4
mAb 4 Protein A pool lot 2	24	≥4.0
mAb 4 Protein A pool lot 3	30	≥4.2
mAb 4 Protein A pool lot 4	64	≥3.9

[a] Inactivation conditions: pH 3.6, 20°C, 30 minutes

The data provided in Table 14 represent the virus inactivation validation for 7 monoclonal antibodies in Protein A pools containing 0.1-0.25M acetic acid, and 5-12 g/L mAb. The inactivation at pH 3.6 at 15°C for 30 minutes showed equally effective inactivation with no mAb effect observed.

Table 14: Reproducible low pH inactivation of X-MuLV using seven proprietary model mAbs. Data courtesy of Qi Chen and Lenore Norling, Genentech, Inc.

Product	**X-MuLV**
mAb 1	>5.6
mAb 2	>6.0
mAb 3	>6.0
mAb 4	>6.2
mAb 5	>6.5
mAb 6	>6.1
mAb 7	>5.1*

*Lower volume tested than other mAbs

In summary, the low pH X-MuLV inactivation is robust, effective and reproducible under the Genentech platform process conditions. The mAb itself, impurity levels, cell viability at the time of harvest, Protein A elution buffer acetic acid molarity and mAb concentration within the range studied do not impact inactivation. The important parameters are pH, temperature, and time. Incubation at pH <3.6, temperature > 15°C and time > 30 minutes are robust and effective inactivation conditions capable of achieving > 6 LRV.

Wyeth *(R. Wright)*

Wyeth presented a data review of 23 studies of A-MuLV and X-MuLV low pH inactivation in Protein A pool (Table 15). Data was analyzed based on protein type, pH, Cl^- concentration, buffer, and study type. Few clear trends among factors were identified. LRV from 13 low pH hold studies ranged from 4.2 to >7.2, while LRV from 10 studies where combined Protein A and low pH inactivation from elution buffer ranged from 5.8 to >8.1.

Table 15: Summary of clearance of X-MuLV by low pH treatment using proprietary mAb process intermediates. Data courtesy of Richard Wright, Wyeth.

Protein class	**Min LRV**	**Max LRV**	**N**		**[Cl-]**	**Min LRV**	**Max LRV**	**N**
RFP[a]	5.7	>5.9	3		0 to 20	4.3	>7.9	6
IgG	4.2	>8.1	16		32 to 54	4.2	>7.6	9
Ig[b]	6.3	>6.9	4		60 to 200	5.7	>8.1	8
Study Type	**Min LRV**	**Max LRV**	**N**		**Buffer**	**Min LRV**	**Max LRV**	**N**
Pro/A pH	5.8	>8.1	10		Acetate	5.7	>5.9	5
Low pH	4.2, 4.3	>7.2	13		Citrate	6.8		1
pH	**Min LRV**	**Max LRV**	**N**		Glycine	4.2, 4.3	>7.9	16
3.8 (pH)	4.2	>7.2	8		Arginine/ HEPES		>8.1	1
3.8 (ProA/pH)	5.75	>7.6	2		**pI**	**Min LRV**	**Max LRV**	**N**
3.9 (pH)	6.92	6.92	1					
4.0 (ProA/pH)	>7.0	>7.5	2		4.5 to 8.0	4.2	>8.1	13
4.1 (pH)	>5.41	>6.8	4					
4.1 (ProA/pH)	5.8	>8.2	6		8.0 to 9.2	4.3	>7.9	10

[a] RFP = Receptor Fc fusion protein

[b] Ig = non-human immunoglobulins that bind protein A but have significantly different structures and molecular weight from human mAbs

In order to generate a dataset to support a generic claim for low pH inactivation, a study was performed to evaluate the effect of various type of products and buffer compositions. Detailed conditions are listed in Table 16 a-c. As shown in Table 16 d, LRV ranged from 4.65 to 6.5 log_{10}. There is no clear effect of buffer and protein type, nor protein concentration. A slight effect of Cl^- concentration in glycine buffer is observed.

Table 16: X-MuLV clearance DoE study using different product types (a), buffers (b), and parameters (c) all studied at worst-case. Reported clearance values (d). HMW: high molecular weight species; POI: product of interest. Data courtesy of Richard Wright, Wyeth.

a) Product	HMW1 (%)	HMW2 (%)	POI (%)	LMW (%)	HCP (ppm)	Protein A (ppm)	pI
Ig[b]	27.5	15.8	56.7	0	5659	5	7.9
RFP[a]	0.5	1.3	98.2	0	1890	2	4.9
mAb	0.2	0.6	97	2.2	5057	4	9.1

b) Buffer species, [NaCl]
50 mM glycine, 18 mM NaCl (20 mM Cl^-)
50 mM glycine, 148 mM NaCl (150 mM Cl^-)
50 mM acetic acid, 20 mM NaCl (20 mM Cl^-)
50 mM acetic acid, 150 mM NaCl (150 mM Cl^-)

c) Parameter	Value
Temp.	18°C
pH	3.8
Time	60’
mAb mg/mL	4-5, 15-25

[a] RFP = Receptor Fc fusion protein

[b] Ig = Non-human immunoglobulins that bind protein A but have significantly different structures and molecular weight from human mAbs

d) Solution characteristics			Protein characteristics					Virus inactivation			
Run	Buffer	[Cl-]	Protein type	[Protein]	pI	HMW	MW	LRV		95% Confidence	
		mM	Protein type	mg/mL		%	kDa	Run 1	Run 2	Run 1	Run 2
1	Acetate	-1	mAb	15-25	9.1	0.5	145	≥ 6.40	≥ 6.70	0.21	0.08
2	Acetate	-1	BDFP	15-25	7.9	42.7	104	≥ 6.15	≥ 6.20	0.05	0.11
3	Glycine	-1	BDFP	15-25	7.9	43.6	104	4.65	5.22	0.18	0.36
4	Glycine	1	RFP	15-25	4.4-5.3	1.9	150	≥ 6.53	≥ 6.18	0.15	0.10
5	Acetate	1	RFP	15-25	4.4-5.3	0.5	150	5.44	≥ 6.78	0.25	0.19
6	Glycine	-1	mAb	15-25	9.1	0.4	145	5.34	4.77	0.30	0.16
7	Acetate	1	mAb	15-25	9.1	0.5	145	6.31	≥ 6.43	0.58	0.15
8	Acetate	-1	RFP	15-25	4.4-5.3	0.6	150	6.05	≥ 6.59	0.44	0.09
9	Glycine	1	BDFP	15-25	7.9	43.6	104	6.14	6.54	0.06	0.33
10	Glycine	-1	RFP	15-25	4.4-5.3	1.9	150	≥ 6.39	6.12	0.06	0.49
11	Glycine	1	mAb	15-25	9.1	0.4	145	≥ 6.55	6.56	0.12	0.41
12	Acetate	1	BDFP	15-25	7.9	42.8	104	≥ 6.15	≥ 6.11	0.21	0.13
13	Glycine	1	BDFP	4-6	7.9	43.5	104	≥ 6.25	≥ 6.39	0.17	0.42
14	Glycine	1	RFP	4-6	4.4-5.3	1.9	150	≥ 6.26	≥ 5.96	0.12	0.17
15	Glycine	1	mAb	4-6	9.1	0.4	145	≥ 6.35	≥ 6.31	0.06	0.40
[Cl^-]: -1 = 15-25 mM											
[Cl^-]: 1 = 140-160 mM						Minimum LRV		4.65			
95% Confidence = 95% confidence interval for LRV values											

a. The symbol (≥) indicates that the output sample contained a virus titer that was below the detection limit of the assay.

BDFP = Binding Domain Fc fusion proteins

RFP = Receptor Fc fusion proteins

Wyeth's historical data and the study evaluating parameter effects support a modular claim for Wyeth's platform processes. Based on this analysis, a modular claim of 4.65 $\log_{10}$ for pH 3.8 or less, 18°C or greater, 60 minutes or more, can be made under the following conditions:

- Glycine and Acetate buffers
- 20 to 150 mM [Cl^-]
- 5 to 25 mg/mL mAb

Novartis *(M. Heitzmann)*

Novartis presented data from 17 studies of mAbs for phase I/II or III trial packages (Table 17). Low pH inactivation claims are based on data from pH 3.6 for 60 min, but 3 time points are routinely measured in the studies. The Protein A pool is in 20-50mM acetic acid/ Tris and is adjusted to the treatment pH with phosphoric acid in an acetate/phosphate buffer system. Table 17 shows the results from duplicate runs performed using large volume plating for sensitive virus testing. The pH 3.6-3.8 treatment was always effective with more than 5 $\log_{10}$ virus infectivity reduction obtained for the 14 different antibodies tested. In conclusion, at least 5 $\log_{10}$ can be assumed for low pH inactivation when the treatment involves a pH ≤ 3.6 and an incubation time of ≥ 60 minutes.

Table 17: Clearance of X-MuLV by low pH using fourteen proprietary mAbs. Data courtesy of Markus Heitzmann, Novartis.

Antibody	**pH**	**X-MuLV[a] run 1**	**X-MuLV[a] run 2**
mAb 1	3.6	>5.8	6.8
mAb 2 study I/II	3.6	6.9	>6.1
mAb 2 study III	3.8	>6.2	>6.6
mAb 3	3.6	6.4	7.0
mAb 4 study I/II	3.6	>7.6	>7.6
mAb 4 study III	3.6	>5.5	>5.7
mAb 5	3.6	>7.1	>7.1
mAb 6	3.6	>7.5	>7.9
mAb 7	3.6	6.7	>5.9
mAb 8 study I/II	3.6	>5.4	>5.3
mAb 8 study III	3.8	>4.6	>4.7
mAb 9	3.6	>6.4	>6.2
mAb 10	3.6	>6.7	>7.1
mAb 11	3.8	>5.5	>5.7
mAb 12	3.6	>4.6	>4.7
mAb 13	3.6	>6.9	>6.7
mAb 14	3.6	>5.8	>5.8

[a] 60 min incubation time

Merck Research Laboratories *(M.E. Dahlgren, N. Tugcu and D.J. Roush)*

Merck reported an anomaly in viral inactivation in low pH process intermediates. The exposure to low pH typically is incorporated into a mAb purification process by holding the Protein A chromatography eluate for a defined time interval. It is then adjusted to neutral pH because the mAb product, like most proteins, has limited stability at low pH. Accordingly, retaining a sample of the low pH process intermediate for subsequent testing of inactivation of a spiked virus necessarily requires extended storage of a process intermediate that is not usually retained for more than a few hours at the most. The low pH might lead to protein degradation over time, and it is possible that degradation products or some other component of this atypical sample matrix might be responsible for the virus inactivation observed in the spiking test. We have documented exactly that anomaly.

While the low pH Protein A product process intermediate tested exhibited over 4 $\log_{10}$ X-MuLV inactivation, the control sample, which was the same material pre-neutralized to pH 7.0 just before the virus spike, exhibited equivalent viral inactivation. Accordingly, it is impossible to defend the claim that the virus inactivation resulted from the low pH effects alone. If the low pH hold test were performed without this process-intermediate control – that is, with simply a control incubation of the virus in culture medium or buffer – this anomaly would not be detected.

The test was repeated with a process stream sample taken from a repeat purification, which was also held at low pH until testing, but for a shorter time period than in the first instance. In this case, the control material that was pre-neutralized immediately prior to virus addition did not inactivate virus. Whatever happened to the process intermediate in the prior case apparently does not occur all the time, but may be time-dependent.

It might be possible to avoid process intermediate degradation by retaining a sample of the neutralized process intermediate, and adding acid to re-adjust it to the elution pH immediately prior to testing. The disadvantage of this approach is that the material tested for virus inactivation is not exactly the same as the process intermediate, inasmuch as it has undergone two rounds of pH adjustment. The most thorough approach would be to retain both a low-pH sample and a neutralized sample. In this scenario, a portion of the sample held at low pH is adjusted to neutral pH immediately prior to testing, and a portion of the material held at neutral pH is adjusted to low pH immediately prior to testing. Each low-pH sample is then tested for viral inactivation, alongside its own neutral-pH control. Inactivation in both cases at low pH, not accompanied by inactivation in the neutral control, makes a convincing case for inactivation attributable solely to the pH effect. We have used this approach successfully in several projects, but it has significant disadvantages in terms of cost.

Regeneron: Case study with low pH hold validation *(J. Mattila)*

Regeneron reported a case study illustrating the importance of process control during low pH hold viral inactivation studies. Regeneron personnel frequently perform external viral validation studies, but the company also has experience outsourcing low pH validation to contract test organizations (CTOs). In this case, an outsourced low pH hold study resulted in X-MuLV inactivation well below expectations. The result was later repeated at the CTO by trained Regeneron personnel. Viral inactivation during placebo controls matched the anomalous test result, strongly indicating that reduced inactivation was not product related.

Their investigation identified the pH meter electrode selection, calibration, and maintenance at the CTO as a likely source of pH measurement error. Virus preparation and mode of acid addition were also investigated but appeared not to contribute to reduced inactivation. Rigorous controls were recommended including:

- Utilization of Tris and protein compatible electrodes
- Broad calibration
- Verification of calibration with standard solutions other than those used for calibration
- Checks for electrode drift with the same standards

Implementation of the new procedure during a subsequent study at the CTO ensured accurate pH determination. Results were obtained in line with all other Regeneron product candidates at comparable conditions, and with the literature [9]. These results emphasize the importance of sponsor process control and process transfer during viral clearance studies at CTOs.

Overall Summary, Low pH Treatment

The data presented for low pH retrovirus inactivation covers a very large dataset. There were 254 low pH studies analyzed from the FDA database and 72 from PEI. Data presented by the 8 companies were from over 85 mAbs and Fc fusion proteins. The majority of the studies, 86% in the FDA database and 87% in the PEI database, showed effective inactivation of MuLV with an LRV range of 4-8 log_{10}, and virus was often inactivated to below the detection limit. The extensive data support that low pH is effective and robust in inactivating retrovirus, although it is important to carefully plan and document technical details concerning the actual validation study. The conclusions from this session fall into four major categories: (1) effects of parameters; (2) understanding outliers; (3) modular validation and company specific platforms and (4) the importance of well-controlled studies.

Effects of parameters

Several companies studied the impact on retroviral inactivation by parameters such as Protein A elution buffer type and concentration, acidification solution, mAb type and pI, protein product concentration, and salt concentration. In general, no clear correlation was found between these parameters and the effectiveness in viral inactivation. Amgen's data suggest that low concentration glycine-containing buffer may slow the inactivation kinetics at pH 3.7 but not at pH 3.6, whereas Wyeth observed reduced inactivation kinetics with low sodium chloride and high protein concentration on inactivation with glycine buffer. Genentech showed that low pH retrovirus inactivation was not impacted in Protein A pools with HCCF added in, suggesting robustness with regard to process and product-related impurity levels.

Data presented by industry include low pH hold from pH 3.5 to 4.1. Although higher pH can still be effective, it is clear that higher pH sometimes correlates with lower LRV, more residual virus detected and slower inactivation kinetics. The data also supports that lower pH, higher temperature and longer time improve low pH viral inactivation. It was agreed that pH $\leq$3.6 is the most robust. For products that are not stable at this pH, it is still possible to obtain effective retroviral inactivation at a pH close to 4, although more variation might be expected.

Information regarding the purity of virus stock used in the low pH studies is generally missing. It is also not clear whether continuous mixing during the low pH hold is used. Both factors can potentially impact the effectiveness of viral inactivation however and so should be further studied.

Understanding the outliers

There were outliers observed in both the FDA and PEI databases where LRVs were lower than expected. PEI described 9 cases; some of them with inactivation significantly lower than 4 LRV. From the information provided in the filing it was difficult to assign a cause; the pH used ranged from 3.7 to 3.8 and inactivation was mostly at room temperature. It is important to investigate these cases further in order to understand why they were not consistent with the general industry experience. For cases where a simple explanation (e.g. malfunction of pH probe) is not available, the sponsor should first investigate whether the data is reproducible and not the result of experimental error. If the effect is real, then further investigation is needed to understand the root cause. The information from these investigations is critical, as it can help identify key parameters for low pH viral inactivation.

In summary, it is critical to ensure that the equipment (e.g. the pH probe) is in good working order, is appropriate for the types of samples being tested, and is calibrated correctly for the pH being tested. Finally, it is important to test the pH of the sample after the virus has been spiked to ensure that the pH is still at the target value.

Modular viral validation and company specific platforms

Several companies reported that their data is consistent with the published parameter ranges and modular LRV value, and proposed to make modular claims for this effective and robust viral inactivation step. While the data from other firms and/or published literature support the robustness of the step, these data should not be used alone to support modular viral validation. In-house data, where all of the process attributes and parameters are thoroughly understood, can provide the complete confidence that a new product/process will clear virus to the same extent as in previous products.

Each company should evaluate its platform conditions for low pH hold, and make modular low pH claims based on its understanding of the impact of process parameters on reproducible inactivation from previously validated products. Hence, it is crucial for companies to troubleshoot any unexpected results to ensure that the same level of viral clearance can be expected when previously validated conditions are applied to a new product.

Importance of well-controlled studies

Two case studies were presented at the symposium where anomalies were observed. In the first case, the process intermediate appeared to inactivate virus at neutral pH. A retest did not reproduce the original result. Although the underlying cause of the anomaly was not identified, it shows the importance of controls to ensure that the reported inactivation was due to the low pH and not the process intermediate. In the second case study, a lower than expected low pH inactivation was obtained. The troubleshooting effort identified an equipment malfunction at the contract testing lab. The subsequent experiments with a more rigorous procedure gave expected results. This case study highlighted the need to have detailed procedures and controls when working with contract testing labs.

Based on the case studies, it was recommended, as one of the next steps, to generate a standard low pH study protocol template with details such as proper study controls, sample handling, and equipment qualification to ensure consistency in the study design and performance. This template could address basic technical questions such as: "was the pH re-verified after virus spike addition", "how was the virus prepared (i.e. crude vs. purified)" and "was the sample mixed or static during the incubation time". Such information will allow clients and regulators to better assess the validity of the scale-down model and viral clearance information gathered at the CTO site. A better understanding of how small-scale modeling is conducted may also address scale-up issues, such as how the minimum incubation time in manufacturing is set, based on the small scale validation data (e.g. the need for extra buffer time).

Miesegaes G, Bailey M, Willkommen H, Chen Q, Roush D, Blümel J, Brorson K (eds): Proceedings of the 2009 Viral Clearance Symposium. Dev Biol (Basel). Basel, Karger, 2010, vol 133, p 43.

SESSION IV

Anion Exchange Chromatography (AEX)

Miesegaes G, Bailey M, Willkommen H, Chen Q, Roush D, Blümel J, Brorson K (eds): Proceedings of the 2009 Viral Clearance Symposium. Dev Biol (Basel). Basel, Karger, 2010, vol 133, pp 45-56.

Anion Exchange Chromatography (AEX)

Technical Background

Anion-exchange chromatography is often used as a polishing step in the purification of mAbs and is typically operated in FT mode [11-13] where the mAb does not bind to the resin while trace impurities (e.g. HCP, DNA, and aggregates) bind to the resin. The same mode of operation is employed for AEX membrane adsorbers [14-17]. Most interactions can be predicted based on electrostatic charge-charge interactions. Operating conditions reported in the literature are typically pH 8 and conductivities less than 7 mS/cm [18]. These conditions are near or slightly below the pI of many mAb isotypes resulting in repulsion of the mAbs from the AEX surface. Under these operating conditions, viruses such as X-MuLV, MMV and SV40 with corresponding pI of 5.8, 6.2 and 5.4, respectively [19], are strongly retained allowing for good virus clearance.

Regulatory Agency Presentation

FDA *(G. Miesegaes)*

We were able to accumulate a substantial number of records of MuLV and MMV clearance by AEX, with a 2:1 ratio in favor of flowthrough (FT) vs. bind and elute (BE) mode (MuLV: n=123 FT mode, n=61 BE mode; MMV: n=41 FT mode, n=28 BE mode). As with Protein A and chemical inactivation records, we noted a shift in the LRV to a lower distribution for those records reporting incomplete vs. complete clearance (data not shown). In this case, the lower LRV may reflect AEX steps run under conditions optimal for protein recovery and separation but not optimal for viral clearance. This may happen, for example, with a product whose pI is similar to that of the study virus.

We first compared the mean LRV for records using infectivity vs. PCR detection methods for both FT and BE records. To perform this analysis, we grouped the MMV and MuLV data; when analyzed separately, similar results were found (data not shown). This grouping was possible because there is no pH-extreme inactivation component for AEX. Unlike that for Protein A (see Fig. 2), we found that AEX records using PCR reported clearance values higher than those using infectivity (Fig. 16a, b). This could be because the increased sensitivity of PCR adds an additional 0.5-1.9 log_{10} to the study LRV window, depending on the mode and application (i.e. BE vs. FT; capture vs. downstream). Alternatively, the additional clearance could be the result of highly efficient capture of viral nucleic acids present in some spike preparations, or even a skewing effect across groups (e.g. if the mAbs in the group where Q-PCR assay is used happen to have higher pIs, leading to better separation of virus from the mAb.

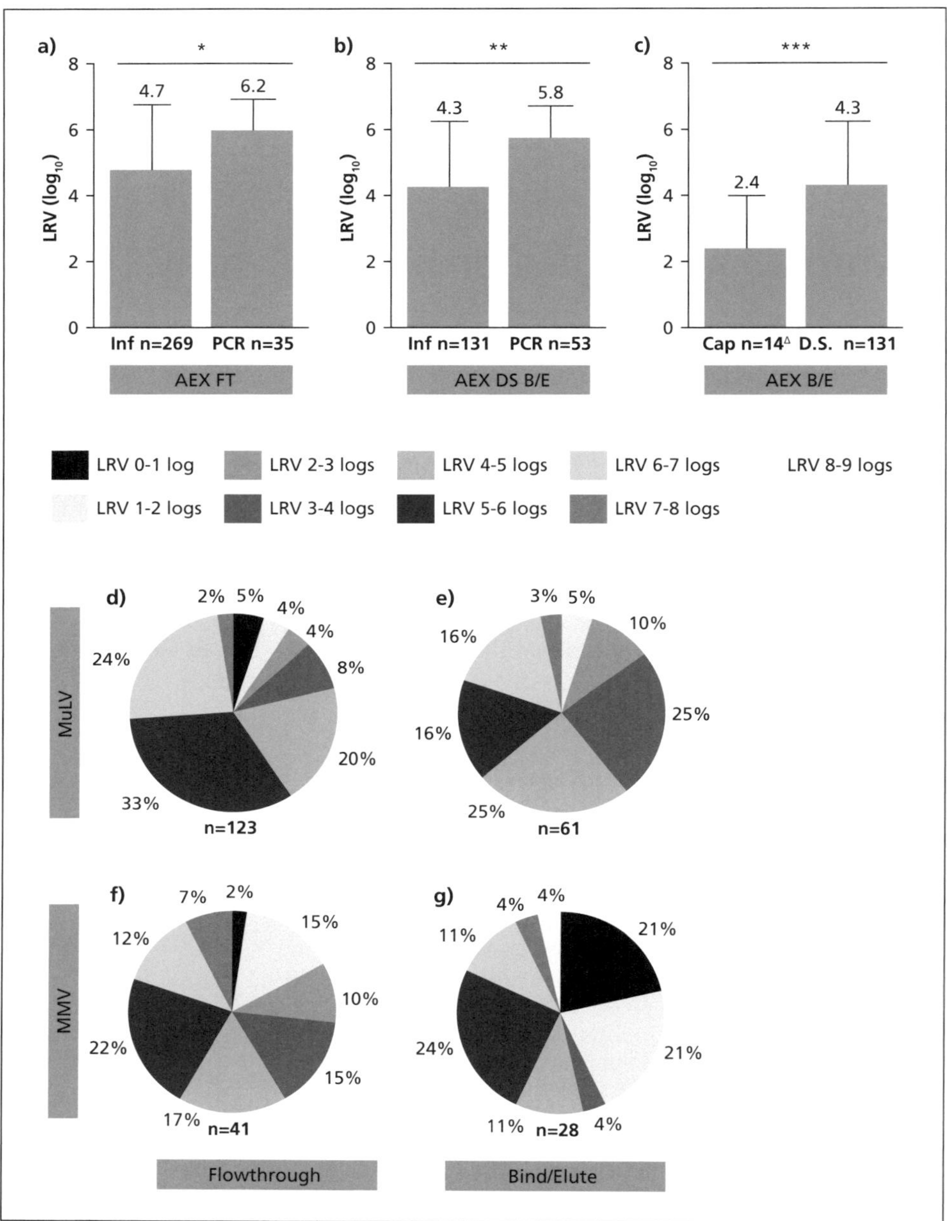

Fig. 16: Analysis of anion exchange chromatography (AEX) records in CDER's viral clearance database. (a,b) Comparison of infectivity (Inf) and PCR-based detection for all viruses (flowthrough mode and bind/elute mode records analyzed separately). (c) Bind/elute AEX used as a capture step vs. further downstream (infectivity records only). Bars represent standard deviations (d-g) LRV ranges binned in 1 log_{10} intervals for flowthrough (d,f) and bind/elute (e,g) modes, MuLV and MMV shown. *p=0.02; **p<0.0001; ***p=0.0007. FT = flowthrough, DS = downstream. △Low number of entries noted. Data courtesy of George Miesegaes, FDA.

Next, we found that when run in BE mode, AEX employed as the first (or capture) step reported LRV almost 2 $\log_{10}$ lower than when run as a polishing step ($p<0.001$; Fig. 16c). This argues that AEX is sensitive to feedstock composition as suggested for Protein A (see Session II). An alternative explanation is that low pI products are more amenable to BE mode AEX (many viruses typically used in viral clearance studies have a relatively low pI) which could result in lower LRVs. It should be noted that the only AEX viral clearance data reported with 8-9 $\log_{10}$ LRV came from BE mode studies, indicating that viral clearance performance of this step can be variable, depending on the process.

We then compared LRV ranges reported for both FT and BE mode AEX (Fig. 16d-g). Both AEX modes appear effective in clearing MuLV (Fig. 16d,e). Only 9% FT and 5% BE records cited LRV <2 $\log_{10}$). In addition, many records cited highly effective values (i.e. in excess of 6 $\log_{10}$) for both modes (25% of FT records and 19% of BE records, respectively). Likewise, AEX can efficiently remove MMV (Fig. 16f,g); 19% of flowthrough and 19% of BE records reported LRV of 6 $\log_{10}$ or more. The LRV distribution for MMV was skewed towards lower values compared to MuLV, which may reflect a slightly lower pI value for MuLV [19]. We did note that almost half of the MMV records using BE AEX (48%) reported LRV of 2 or less (Fig. 16g), none of which claiming complete clearance. In all likelihood, this is not a limitation of the technology but rather the result of protein product pIs being in the acidic range and near to the pI of MMV.

Industry Presentations

Presentations at the Symposium confirmed previous data from the literature on operating conditions for AEX that achieve robust LRV for X-MuLV [18]. This is of interest as X-MuLV is the virus cited in ICH Q5A risk assessment [20]. Moreover, the presentations extended these observations to 27 mAbs (each with various properties) run under a range of operating conditions (pH, conductivity, etc.) and resin types.

However, several key questions remained to be addressed including the following: (1) potential impact of resin loading on AEX LRV; (2) impact of residence time on AEX LRV (comparison of membrane vs. packed bed); and 3) whether any general trends in viral retention on AEX across viruses and resins existed (e.g. is the combination of virus pI and operating conditions, including pH, conductivity, and buffer type generally predictive of retention?). These topics were debated in detail at the Viral Clearance Symposium.

General observations conserved across AEX studies

When resin type is conserved and AEX is operated in FT mode there appears to be a correlation between decreased LRV for X-MuLV and the conductivity of load material (Fig. 17; Human Genome Sciences) when it is greater than or equal to 15 mS/cm. Additional data presented by another firm confirm the insensitivity of LRV for X-MuLV to conductivity up to 14 mS/cm (Fig. 18; Boehringer Ingelheim) with an LRV of >4 $\log_{10}$ achieved for all conditions up to this conductivity. Salt tolerant or mixed-mode ligands and Q-membrane adsorption offer the potential to increase the operating range to conductivities > 14 mS/cm [21].

Experimental results from a third firm (Table 18; Abbott Bioresearch Center) obtained over a range of buffer conductivities (5 to 10 mS/cm) at a fixed pH (8.0) suggest that this observation of insensitivity to conductivity may be conserved for both X-MuLV and MMV. Throughout the entire parameter space evaluated, high levels of viral clearance (> 4 LRV) were achieved for four mAbs tested (pI range: 8.4 – 9.4). Note that the apparent insensitivity of LRV seen by all three firms to conductivity below 10-14 mS/cm is similar to the results from Curtis et al (2003) on the impact of pH and conductivity on SV40 clearance by AEX [22, 23].

Clearance studies with a Q-membrane adsorber from another firm (Table 19; MacroGenics) support insensitivity to even higher conductivity (range 8-17 mS/cm, pH 7.4-8.4). Over 5 LRV was achieved for MuLV under all conditions and for PPV at pH 8.3-8.4 (D. Farb – MacroGenics). Perhaps due to the higher pI of PPV, only 3-4 LRV was observed when PPV was loaded on Q-membranes in two separate antibody studies at pH 7.4 and 16-17 mS/cm.

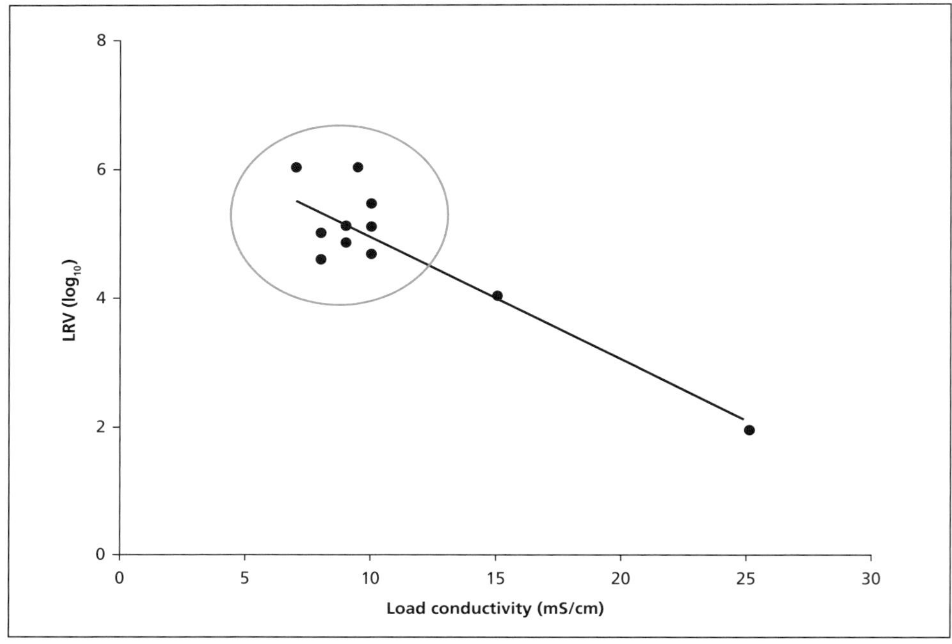

Fig. 17: Clearance (infectivity) of X-MuLV by flowthrough AEX using six undisclosed proprietary mAbs, vs. conductivity (measured by Q-PCR). Data courtesy of Olga Galperina, Yaling Wu, David Kahn, and Yuling Li, Human Genome Sciences (HGS).

Table 18: Clearance of X-MuLV, MMV, Reo-3, and PRV by flowthrough AEX chromatography using four proprietary mAbs. Data courtesy of Edwin O. Lundell, Abbott Bioresearch Center.

mAb	Matrix	Load g/L resin	X-MuLV	MMV	Reo-3	PRV
A	Tris/acetate/PBS pH 8, conductivity 5-10 mS/cm	50	≥5.49 ≥4.88	≥5.57 ≥5.30	ND	ND
	Tris/acetate/phosphate pH 8, conductivity 5-10 mS/cm	43 76 43	≥4.42 ≥4.85 ≥4.55[a]	4.14 4.34 3.10[a]	3.34 4.51 3.99[a]	≥4.80 ≥4.80 ≥4.86[a]
B	Trolamine/PBS pH 7.7, conductivity 5-10 mS/cm	60	≥5.51 ≥5.59	≥7.04 ≥7.04	ND	ND
C	Tris/acetate/phosphate pH 8, conductivity 5-10 mS/cm	30	≥6.21 ≥6.21	≥8.40 ≥8.40	ND	ND
D	Tris/acetate/phosphate pH 8, conductivity 5-10 mS/cm	69	≥6.49 ≥6.48	6.73 7.71	ND	ND

[a] Aged resin (50 cycles)

Table 19: Clearance of X-MuLV, Reo-3, and PPV by membrane adsorption using two proprietary mAbs. Virus Spike Study Conditions, 0.7 minute residence time. Data courtesy of David Farb, MacroGenics.

Process step Q-membrane anion adsorption	**Virus spike study conditions**				**Virus infectivity $\log_{10}$ reduction**		
	g/L conc.	pH	mS/cm	g/L load	X-MuLV	Reo-3	PPV
mAb2 (pI~8.6)	3.6	7.4	17.2	373	> 5.72	n.d.	4.03
	2.6	8.4	16.3	262	> 5.59		6.00
mAb1 (pI~9.1)	6.2	7.5	15.6	1331	> 5.65	> 5.39	3.36
	6.2	8.3	7.9	1317	5.49	> 5.25	> 5.66

Other studies demonstrated that robust clearance of X-MuLV (>5 Log_{10}) could be consistently obtained using AEX (Q Sepharose FF) in the flowthrough mode at a pH as low as 6.0 for a variety of antibodies using a very low conductivity buffer (< 5 ms/cm). However, using the same conditions, MMV clearance was more modest, ranging from 3-5 LRV using a pH of 6.0-6.2, but increased to > 5 LRV when operating at a pH ≥ 7.0 (M. Barry – ImClone). These differences may also be related to the slightly higher pI for MMV relative to X-MuLV.

The combination of the aforementioned observations suggests that electrostatic interactions dominate the differential retention of X-MuLV (pI of 5.8) on AEX vs. the mAbs. Once the conductivity is raised above a threshold, selective retention of the virus relative to the mAb is lost. Additional data (reference Table 20) obtained via use of AEX in bind/elute mode (inverse mode to typical AEX flowthrough) confirm this idea. Efficient retention of MuLV or MMV resulting in high LRV was obtained when the antibody could be eluted at pH > 6.8 at low salt strength/conductivity. Two antibodies with comparatively low pIs required a higher salt strength for elution. In these cases, lower LRVs were obtained when the peak-cut window for the antibody exceeded 14 mS/cm, which is indicative of virus elution above this conductivity value.

Table 20: Clearance of X-MuLV and MMV (LRV) by flowthrough AEX using twelve proprietary mAbs (range of pI 6.2 to 9.1). Data courtesy of Markus Heitzmann, Novartis.

Antibody	**pI**	**Elution mode /conditions**	**X-MuLV**	**MMV**
mAb 2	8.7	pH step /pH 7.8 < 5mS/cm	> 6.1	> 8.1
mAb 3	8.5	pH step /pH 6.8 < 5mS/cm	> 6.8	not tested
mAb 4	8.5	pH step / pH 7.8 < 5mS/cm	> 6.2	> 7.5
mAb 5	9.1	pH step / pH 7.8 < 5mS/cm	> 5.7	> 6.3
mAb 6	8.9	pH step / pH 6.8 < 5mS/cm	> 7.0	not tested
mAb 7	8.9	pH step / pH 7.8 < 5mS/cm	> 6.1	not tested
mAb 8	8.9	pH step /pH 6.8 < 5mS/cm	> 7.5	not tested
mAb 9	8.5	pH step /pH 6.8 < 5mS/cm	> 7.0	> 6.1
mAb 13	7.7	pH step /pH 6.8 < 5mS/cm	> 6.2	8.1
mAb 10	6.6	pH step /pH 6.8 < 5mS/cm	> 6.9	4.9
mAb 12	6.4	Salt gradient /pH 7.8, 5-11 mS/cm peak-cut window	> 5.0	> 7.2
mAb 14	6.2	Salt gradient /pH 7.8, 6-14 mS/cm peak-cut window	-	> 6.9
		Salt gradient /pH 7.8, 14-20 mS/cm peak-cut window	-	3.2
		Salt gradient /pH 7.8, 12-28 mS/cm peak-cut window	1.7	< 1.0

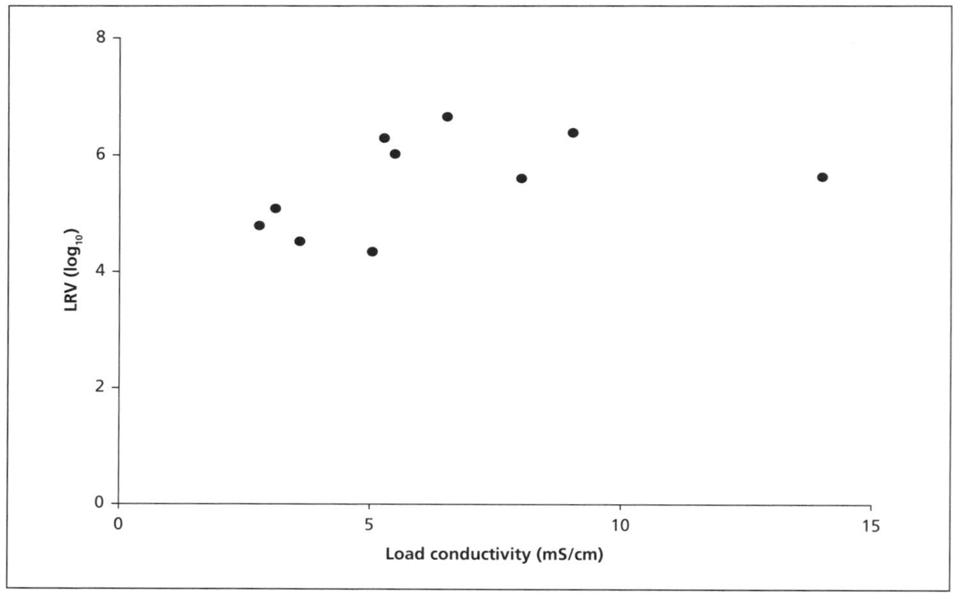

Fig. 18: Clearance of X-MuLV by flowthrough AEX using undisclosed proprietary mAbs, vs. conductivity (measured by infectivity). Data courtesy of Franz Nothelfer, H. Winter and D. Ambrosius, Boehringer Ingelheim Pharma GmbH.

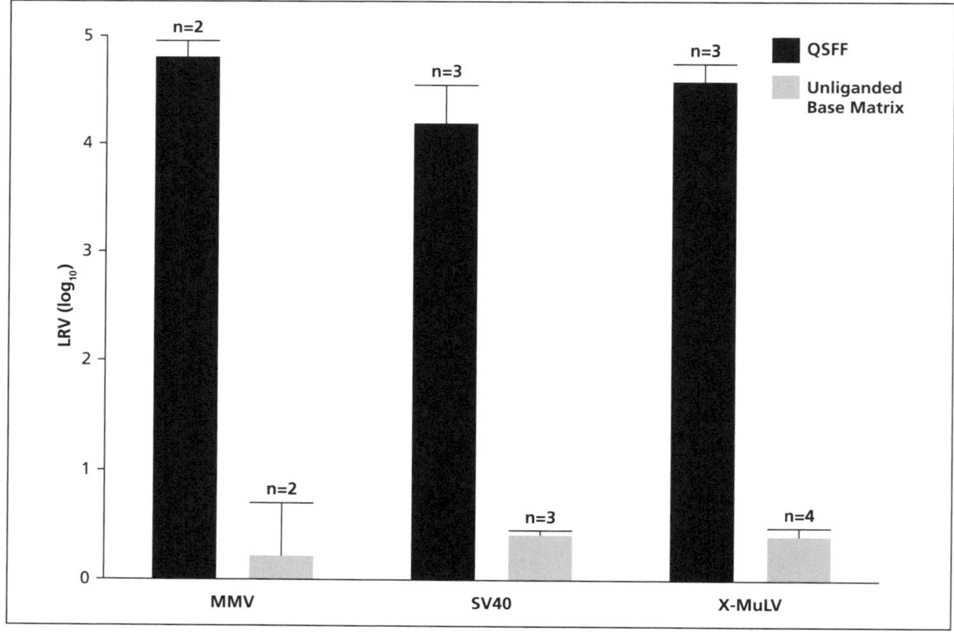

Fig. 19: PCR measured clearance of X-MuLV, MMV, and SV40 by flowthrough AEX using mAbs as described in [23]. Data courtesy of Qi Chen and Lenore Norling, Genentech, Inc.

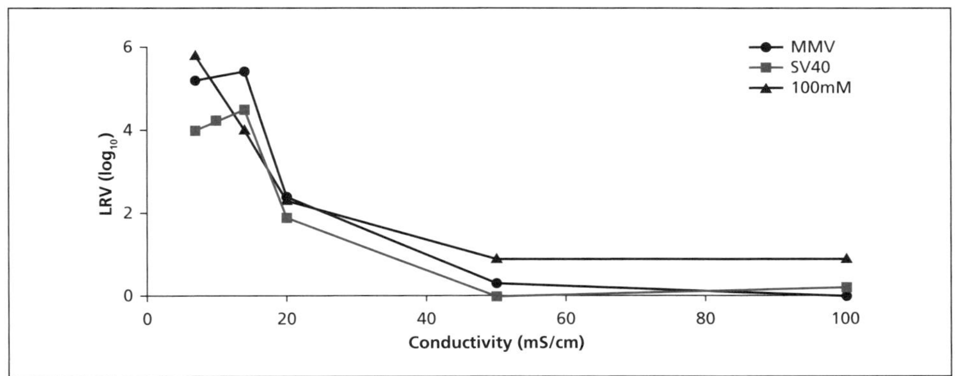

Fig. 20: PCR measured clearance of X-MuLV, MMV, and SV40 by flowthrough AEX as a function of conductivity using mAbs as described in [23]. Data courtesy of Qi Chen and Lenore Norling, Genentech, Inc.

Further supporting evidence for the electrostatic nature of the interaction between viruses (MMV, SV40 and X-MuLV) and the QSFF resin is derived from experiments (Figs 18-20; [23]) demonstrating the significant difference in the LRV levels achieved utilizing the base matrix resin (unliganded) vs. the standard (QSFF) AEX resin. These data clearly indicate that the viral clearance (expressed as LRV) requires the presence of the positively charged ligand (Q in this case) on the resin.

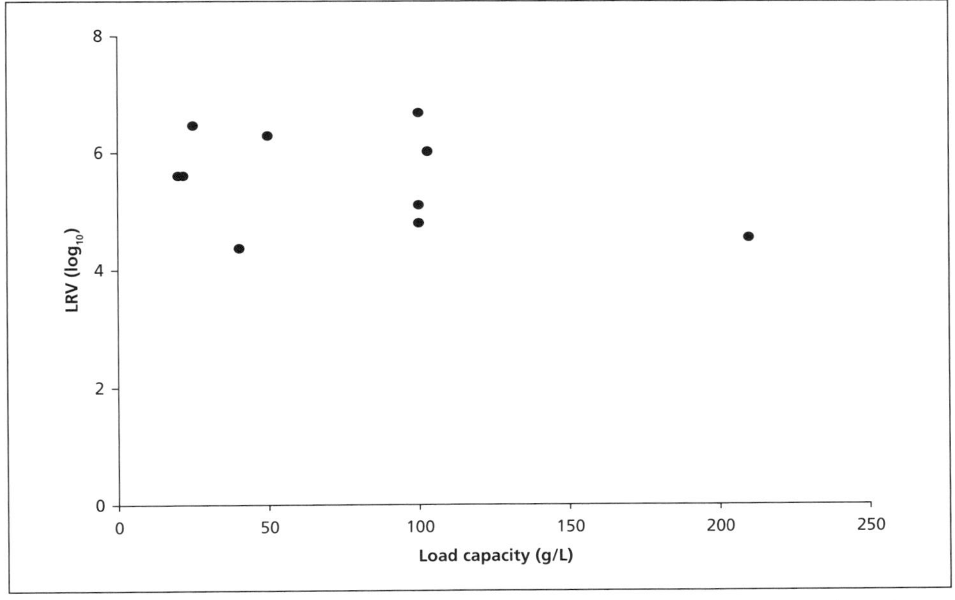

Fig. 21: Clearance of X-MuLV (measured by infectivity) by flowthrough AEX using undisclosed proprietary mAbs, vs. effective loading (mg/mL). Note insensitivity to effective loading up to 200 mg/mL. Data courtesy of Franz Nothelfer, Helmut Winter and Dorothee Ambrosius, Boehringer Ingelheim Pharma GmbH.

A second observation is that X-MuLV clearance (LRV) appears to be insensitive to effective protein loading (AEX operated in flowthrough mode) up 100 mg/mL-resin (Fig. 21). The observation of insensitivity of LRV to effective resin loading was confirmed by others (M. E. Dahlgren – Merck). E. Lundell (Abbott), M.E. Dahlgren (Merck), and P. Mensah (Pfizer) evaluated three different resins (Q Sepharose FF, UnoSphere Q and Capto Q) and suggested that LRV clearance for X-MuLV may be insensitive to the type of AEX resin and effective loading (up to 76 mg/mL) if other key parameters (pH and conductivity) are constrained. Another study demonstrated that good clearance was achieved (> 4 LRV) for other virus models (MMV and BVDV) – clearance that was insensitive to the number of cycles up to at least 50 and that X-MuLV clearance (> 5 LRV) was insensitive up to at least 70 cycles (M. Barry - ImClone, data not shown). However it is important to note that material limitations may dictate the maximum number of cycles since no significant drop in LRV was observed as a function of cycle number for the data presented.

One important point raised during the symposium was the potential for host cell residuals to compete with virus partitioning on the AEX resin, and hence limiting the range of insensitivity to LRV with respect to loading. Experimental data from two presentations touch upon this issue. First, data shown in Table 21 suggests that although LRV is insensitive to effective loading up to 100 mg/mL, increasing the loading to 200 mg/mL may result in a decreased LRV for some process streams. The LRV data in Table 21 (except as noted) were generated using feed to the AEX step (performed with Capto Q), which was the second stage of a given purification process. Viral clearance studies were then performed under the same conditions but with using a fully purified mAb (Table 21); this work suggested that the fully purified mAb was insensitive to loading up to 200 mg/mL. In a second presentation, the potential for host cell residuals (DNA in this example) to impact LRV for X-MuLV was further exemplified for a common AEX resin, Q-Sepharose FF (Fig. 22). A similar trend was not observed for Q-Membrane adsorber studies, where over 5 LRV for MuLV, Reo-3 and PPV was achieved after loading up to 1400 g IgG per L membrane (D. Farb - MacroGenics). Nevertheless, when virus clearance is performed under high IgG load, the potential impact of residual host impurities should be considered.

Table 21: Clearance (LRV) of X-MuLV by flowthrough AEX as a function of load (mg/mL) using three proprietary mAbs (measured by infectivity). Data courtesy of Paul Mensah, Mark Gustafson, and Downstream Purification colleagues, Pfizer.

MAb	Effective loading (mg/mL)	X-MuLV
A	100	>6.63
	200	>6.33
B	100	5.46
	200	2.93
C	100	>5.61
	200	1.4
C (Fully purified)	100	>5.44
	200	>5.13

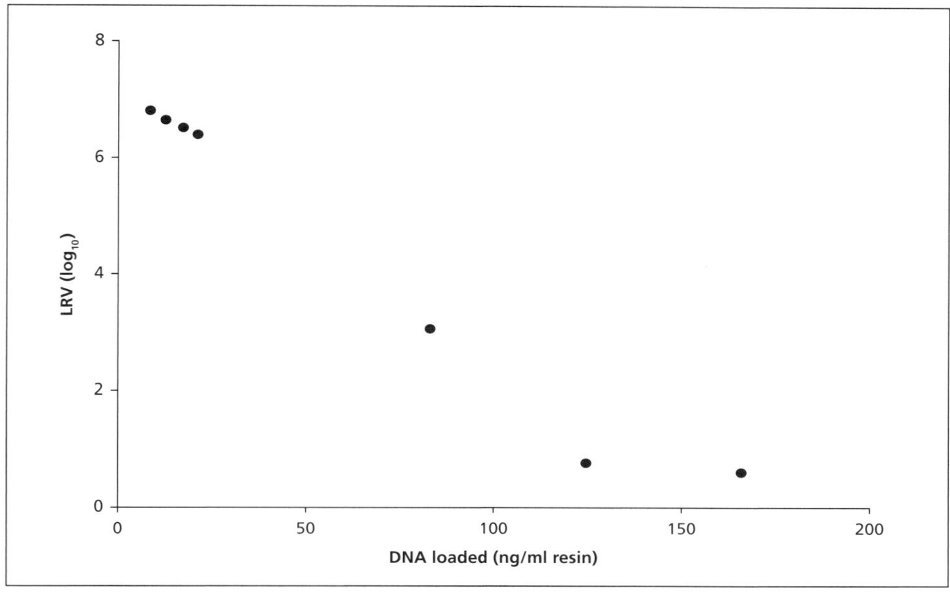

Fig. 22: Clearance of X-MuLV by flowthrough AEX using undisclosed proprietary mAbs, vs. DNA loaded (ng/mL resin) indicating the potential impact of host cell residuals on LRV. Data courtesy of Paul Mensah, Pfizer.

Another interesting observation was that the viral clearance LRV appears to be insensitive to combination of high linear velocity ("High"; maximum of 250 cm/hr) and high effective loadings (up to 150 mg/mL) for four model viruses (Table 22). The insensitivity of viral clearance to linear velocity/short residence times is even more pronounced for the Q-membrane adsorber. In this case, residence time was reduced to 0.7 minutes but still achieved >5 log_{10} clearance of MuLV, Reo-3 and PPV (D. Farb – MacroGenics; Table 19). One important note is that using the Q-membrane adsorber, Polio virus (a neutral pI virus) was not effectively cleared effectively, even under non-worst-case conditions (Table 22).

Table 22: Clearance (LRV) of ERV, PRV, Reo-3, and Poliovirus (Polio) by flowthrough AEX (measured by infectivity) as a function of linear velocity using two proprietary mAbs at 150 mg/mL load density. Data courtesy of Pedro Alfonso, Centocor Ortho Biotech.

Product	Flow-rate	ERV	PRV	Reo	Polio
A	High	> 7.3	> 5.7	> 6.2	0
B	High	> 7.3	> 5.8	> 6.0	0
A	Low	> 7.4	> 6.0	> 5.5	0
B	Low	> 7.3	> 6.0	> 5.6	0.2

A key area of discussion at the meeting was a further delineation of the potential impact of buffering species that contribute to the observed conductivity. Experimental results presented in Table 23 suggest that for a wide range of mAb and Fc fusion proteins, robust levels of LRV can be achieved with a variety of buffer compositions (e.g. glycine and Imidazole, glycine and Tris, glycine and HEPES). Additional experiments are warranted to further extend the bracketed viral clearance results observed using standard systems (e.g. phosphate buffered saline and Tris, acetate and Tris, citrate and Tris) to new resin and buffer combinations.

Table 23: Clearance of X-MuLV by flowthrough AEX (TMAE resin) as a function of buffer composition using a range of undisclosed proprietary mAbs. Data courtesy of Richard Wright, Wyeth.

Protein type	pH	Buffer	~[Cl⁻]	LRV	pI
mAb	7	Gly[b]/ Imidazole	35	>4.29	7.54
mAb	7.5	Arginine/ HEPES	100	>4.34	6.94
mAb	7.5	Gly/ Tris	20	>5.4	8.29
mAb	7.8	Gly/ Tris	50	>5.67	7.99
mAb	8	Gly/ Tris	40	>3.75	7.98
Ig[a]	7.2	Gly/ HEPES	60	>5.30	8.59
mAb	7.2	Gly/ Tris	43	>6.13	6.94
Ig	7.9	Gly/ Tris	54	>6.47	7.54
mAb	7.2	Gly/ Tris	20	>6.52	8.8
Ig	7.9	Gly/ Tris	54	>6.65	9.24
mAb	7.2	Gly/ Tris	20	>4.32	8.87
mAb	7.1	Gly/ Tris	60	>6.25	7.1
mAb	7	Gly/ Tris	60	>5.44	6.97

[a] Ig = non-human immunoglobulins. They bind protein A but have significantly different structures and molecular weight from human mAbs

[b] Gly = glycine.

Extension to other viruses

As summarized above, robust levels of LRV for both X-MuLV and MMV are possible for AEX when operated in flowthrough mode. One pending question is whether the same electrostatic mechanism of binding pertains to other model viruses typically evaluated (including SV40, PPV, Reo-3 and PRV). Data presented in Table 18 and Figure 19 support this hypothesis. Additional experimental results reviewed at the Symposium (Fig. 20) clearly indicate that viral clearance is a function of conductivity for levels in excess of 14 mS/cm for X-MuLV, MMV and SV40 for a specific mAb evaluated.

An observation consistent throughout all of the AEX presentations at the Symposium was that good clearance (> 3 LRV) was achieved for a total of seven viruses – X-MuLV, MMV, PPV, SV40, Reo-3, PRV, and ERV, over a wide range of operating conditions (e.g. see Tables 18, 19, 22). The only exception was Poliovirus (as mentioned above) where poor clearance was observed – an observation which would be consistent with an electrostatic retention mechanism, since the pI of polio is ~8 and is similar to the pI of the mAbs investigated. Taken together, the data presented at the Symposium, along with the literature precedent [22-24], support the concept that a bracketed or generic viral clearance approach (e.g. fixed buffer composition, conductivity, pH and maximum effective loading) is feasible for AEX chromatography operated in flowthrough mode.

Additional experiments were proposed to further characterize the importance of electrostatic interactions (i.e. the differential between the pI of the virus and the mAb). These experiments could potentially include DoE studies (Fig. 23) to (1) further define the LRV response surface for a range of virus types; this information would extend observations seen with X-MuLV to other model viruses; and (2) to better understand the mechanisms of AEX operation in BE mode. In addition, experiments to accurately determine the pI of other model viruses (BVDV, Reo-3, PRV, Polio, PPV) would be desirable, extending the research of Strauss et al [19].

AEX (Q Sepharose FF) is able to provide effective removal of three model viruses, including X-MuLV, a model virus for CHO retrovirus-like particles (RVLP). Genentech determined the ability of a typical QSFF process to remove actual CHO RVLP impurities. Using small scale experiments with three model antibodies, Genentech observed that the typical flowthrough QSFF process is capable of effectively removing both in-process and spiked RVLPs from different feedstocks containing different mAb products [25]. In addition, it was demonstrated that this AEX process achieves a similarly high degree of RVLP removal during large scale manufacturing operations.

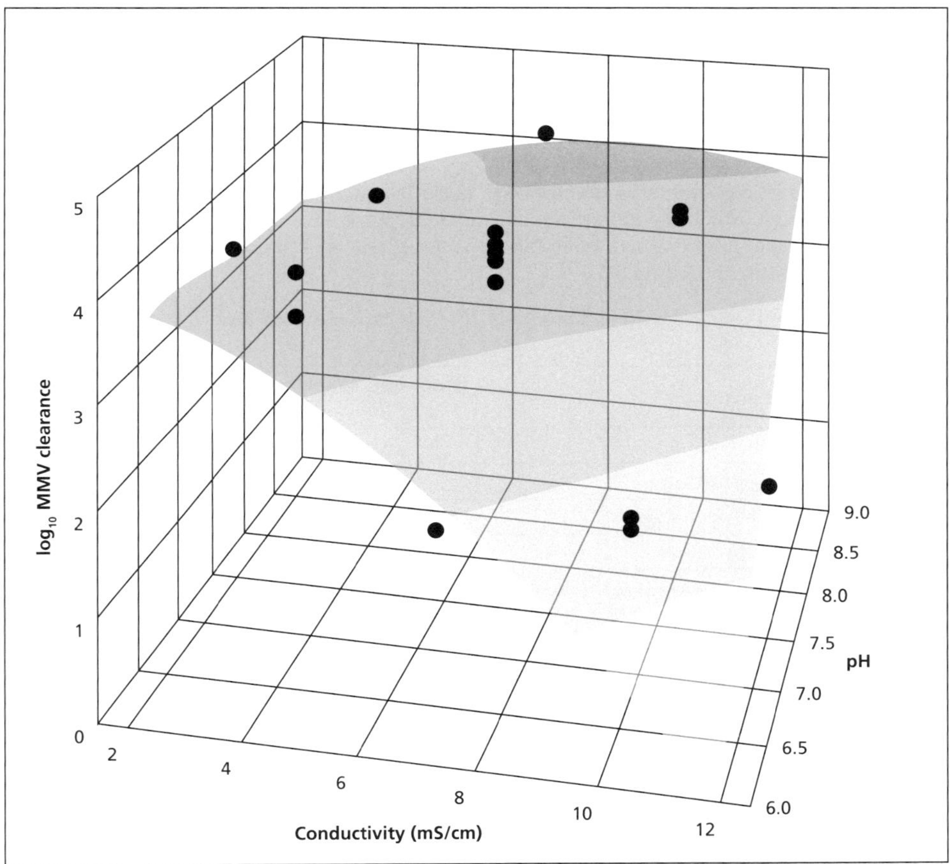

Fig. 23: Response surface for LRV of MMV vs. conductivity and pH (independent of loading). Data courtesy of Mary E. Dahlgren, Merck Research Laboratories.

Overall Summary (AEX)

The group concurred that determining the pI for additional viruses (BVDV, Reo-3, PRV, Polio, and PPV) is important to confirm that electrostatic interactions are the dominant factor impacting the differential binding of antibodies and viruses for AEX. This is an indirect approach and would complement efforts to determine the impact of changes to pH and/or conductivity on LRV. These data would extend the observation from X-MuLV and MMV that one of the key drivers for retention is the difference in the net charge (pI at a fixed pH) of the virus and the antibody. If electrostatic mechanisms are the key drivers for achieving the targeted LRV for the virus then constraining the pH and conductivity should permit robust removal. It should be understood that there will be a subset of viruses where FT AEX will not be an effective virus removal step (i.e. neutral pI viruses such as Polio).

Although viral clearance by AEX appears to be electrostatically driven, there may be secondary parameters that influence clearance, including competitive adsorption by other impurities (e.g. DNA). Further investigation of the impact of buffer and resin combinations is warranted to determine the impact of operating conditions on viral clearance of MMV.

AEX is generally held to provide consistent viral clearance for most model viruses used by industry in validation studies (for a list of the most common validation viruses see [3]) under the proposed bracketed operation conditions (Table 24), with the exception of poliovirus (pI of ~8.0). Although effective viral clearance is feasible with AEX in the bind and elute mode, this step is more commonly operated in the flowthrough mode and is consistently robust (less sensitive to operating conditions).

Table 24: Summary of proposed bracketed conditions (X-MuLV clearance insensitive) to achieve robust viral clearance*

Parameter*	Proposed range for bracketed generic viral clearance
pH	7.0 – 8.5
Conductivity	< 14 mS/cm
Effective Loading	< 100 mg/mL[1]
Resin reuses	50[2]

* These parameters would need to be constrained in order to achieve robust viral clearance. Other parameters (ex. linear velocity) have not been reported to impact LRV.

[1] Effective loading could be higher depending on host cell impurity load (ex. host cell DNA)

[2] Maximum number of reuses could be expanded with additional data. No LRV decrease with increased cycle number could be extended up to 200 cycles if comparable performance achieved to SV40 [11]

Based on the above observations and the scientific literature, the group identified "the defining of a modular viral clearance design space for antibodies with known characteristics" (e.g. pI range) as a high priority for follow-up investigation (Table 24). This may involve additional experimentation to confirm that these "bracketed" operating conditions are conserved across multiple types of AEX resin and operating conditions (e.g. high linear velocity). As indicated above, these experiments could include DoE studies (Fig. 23) to further define the LRV response surface for virus types. Moreover, the determination of virus pIs either through experimentation or theoretical means [26] would inform the design of these experiments. Last, these experiments could also expand the evaluation of the impact of buffer composition on LRV for the various viruses. Data published to date [23] suggest that the LRV for X-MuLV is unaffected by the presence of NaCl or NaAcetate as the buffer salt.

Miesegaes G, Bailey M, Willkommen H, Chen Q, Roush D, Blümel J, Brorson K (eds): Proceedings of the 2009 Viral Clearance Symposium. Dev Biol (Basel). Basel, Karger, 2010, vol 133, p 57.

SESSION V

Alternative Chromatographic Methods

Miesegaes G, Bailey M, Willkommen H, Chen Q, Roush D, Blümel J, Brorson K (eds): Proceedings of the 2009 Viral Clearance Symposium. Dev Biol (Basel). Basel, Karger, 2010, vol 133, pp 59-66.

Alternative Chromatographic Methods

Technical Background

Protein A Chromatography and AEX Chromatography are mostly used in a platform technology to produce monoclonal antibodies. Other chromatographic methods are commonly involved for further purification to achieve a sufficiently low level of HCP, HMW proteins, or DNA in the final product. CEX, HIC, and HA are alternative methods which can be used if the product properties require a specific purification process. The operational range for achieving virus removal by these resins seems smaller than those shown for AEX. Even if virus binding depends on charge and hydrophobicity of the virus membrane, it is less predictive, and the effects are more virus-dependent.

Other media have been developed which have a multifunctional linker. For instance, CaptoAdhere is a strong anion exchanger with multimodal functionality. Most pronounced are ionic interaction, hydrogen bonding and hydrophobic interactions.

Virus removal by these chromatographic methods differs in their mode of action but they have some capacity for virus removal. Some of these methods, such as CEX, are already used for production of a large number of mAbs; others are not.

Regulatory Agency Presentation

FDA *(G. Miesegaes)*

Cation exchange (CEX) is predominantly employed in bind/elute mode to purify antibodies, which tend to be slightly basic proteins. This was reflected in the number of FT vs. BE records we obtained; 88% of MuLV records and 96% of MMV records were BE (capture or polishing). Because of the skewed representation we restricted our analysis to BE mode CEX. We compared the mean LRV obtained from records employing CEX as a capture vs. polishing step (Fig. 24a). To perform this analysis, we grouped MuLV and MMV data, as CEX has only a minimal low pH inactivation component. Similar to AEX, we found that clearance was significantly lower when CEX is used as a capture step (p=0.01). This suggests that ion exchange chromatography is more susceptible to feedstock complexity, relative to Protein A chromatography. This likely reflects differences in mechanism of action; affinity chromatography is more specific for particular target molecules vs. ion exchange chromatography. Thus, interactive effects between affinity-based resins and extraneous cell culture components might be reduced, relative to ion exchange resins.

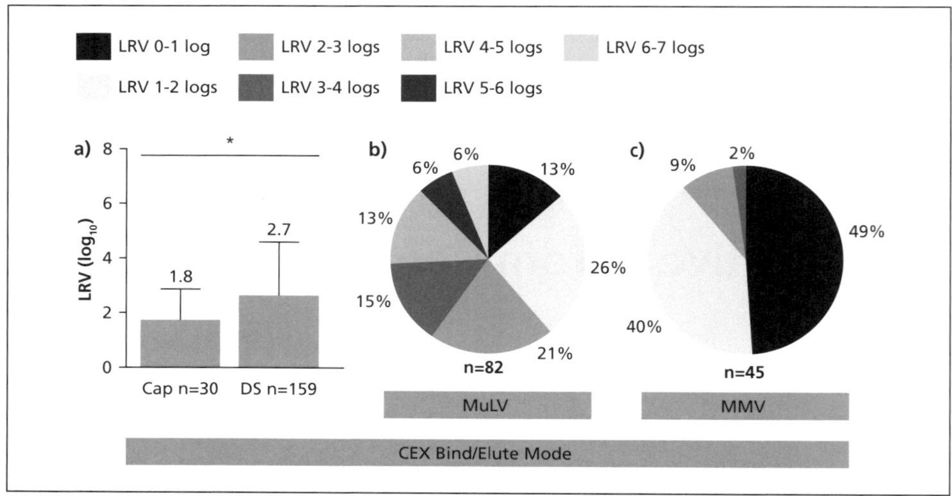

Fig. 24: Analysis of cation exchange (CEX) records in CDER's viral clearance database, bind/elute mode shown. (a) Bind/elute AEX used as a capture step vs. further downstream. Bars are standard deviations. (b,c) LRV ranges binned in 1 log_{10} intervals for MuLV (b) and MMV (c). *p=0.01. Cap capture, DS downstream. Data courtesy of George Miesegaes, FDA.

Clearance by CEX BE mode is more variable than other unit operations (Fig. 24b) [3] and MuLV is generally more successfully cleared than MMV (compare Fig. 24b with 24c). Whereas 40% MuLV records achieved clearance of > 3 log_{10}, only 2% of the MMV records reported a similar level of clearance. In addition, only 39% of the MuLV records cited very poor (<2 log_{10}) removal, compared to a very high (89%) number of the MMV records. Given the poor performance of CEX, it is not surprising that we did not find any MMV records reporting complete clearance values. We concluded based on this data that CEX is only modestly successful in clearing MuLV, and is rather inefficient for clearing MMV. This difference may be due to a pH effect; CEX was found generally to be run using slightly acidic buffers [3] which in theory may inactivate enveloped viruses such as MuLV. This effect would be independent of the actual separation power of CEX, which might be quite poor given the results shown for MMV. MMV also has a slightly higher pI than MuLV [19] and therefore may be more difficult to separate it from neutral or basic pI products by this particular charge based chromatography method. It should also be understood that this is a general conclusion across a wide number of processes and conditions, and does not imply that CEX is ineffective in all situations.

Industry Presentations

Cation Exchange Chromatography (CEX)

Amgen *(L. Connell-Crowley)*

CEX chromatography, often used as a polishing step in mAb manufacturing, was investigated in the binding/elution mode for removal of X-MuLV, MMV, PRV, and Reo-3. Data from several Amgen mAb processes indicated that a CEX resin, Fractogel SO_3^-, can achieve high levels of X-MuLV, PRV, and Reo-3 clearance at pH 5.0 (Table 25). Clearance

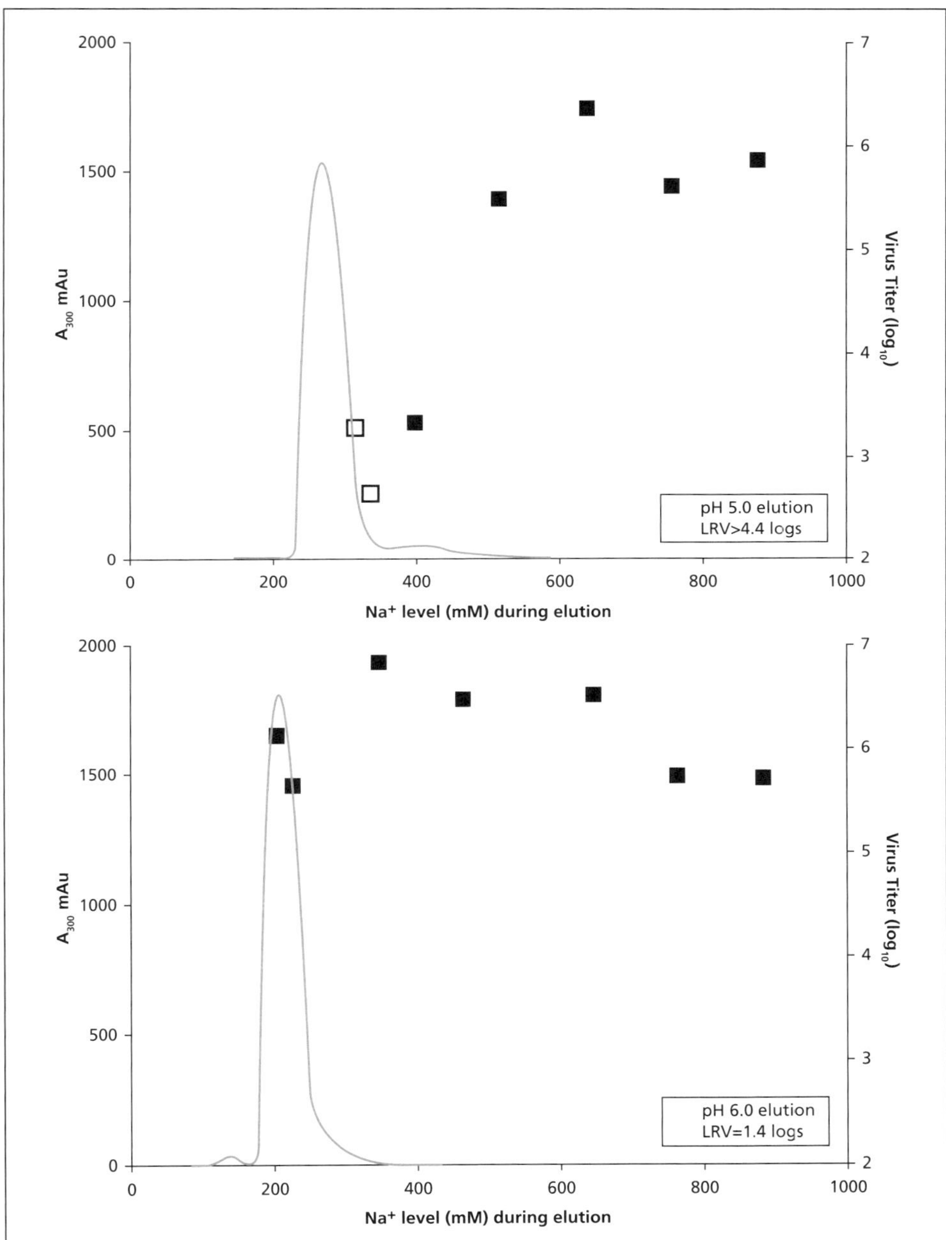

Fig. 25: Clearance (infectivity) of X-MuLV by bind and elute CEX using an undisclosed proprietary mAb. A Protein A-purified mAb was spiked with X-MuLV and loaded onto a Fractogel SO_3- column at pH 5.0 or 6.0. The column was washed and then eluted using a salt gradient at the same pH. Fractions were collected during the gradient and assayed for X-MuLV by $TCID_{50}$. The mAb elution profile is shown as an A_{300} trace (line). The X-MuLV titer of each fraction is shown as squares; open squares indicate the virus level is below limit of detection. Data courtesy of Lisa Connell-Crowley, Amgen.

of X-MuLV and Reo-3 was reduced at pH above 5.0, while in one instance of PRV clearance remains high. In contrast, Fractogel SO_3^- was not able to effectively clear MMV at any pH.

Effective separation of X-MuLV from product occurs because the virus binds to the resin more tightly than the product at pH 5.0. Figure 25 shows a comparison of the elution profiles of a Protein A-purified mAb and X-MuLV eluted from Fractogel SO_3^- using a salt gradient at pH 5.0 and 6.0. At pH 5.0, X-MuLV elutes later in the gradient and is well separated from the product. At pH 6.0, the virus and mAb co-elute, resulting in poor virus clearance. Similar experiments were used to further define an operating window of < 350mM sodium at pH 5.0 for robust X-MuLV clearance. This allows for design of additional X-MuLV clearance into a process that could potentially be validated using a generic approach. Similar experiments to define viral clearance operating spaces can be performed with PRV and Reo-3.

Table 25: Clearance of X-MuLV, MMV, PRV, and Reo-3 (infectivity) by bind and elute mode CEX using thirteen proprietary mAbs. Data courtesy of Lisa Connell-Crowley, Amgen.

Operating pH	Molecule[a]	LRV			
		X-MuLV	MMV	PRV	Reo-3
5.0	mAb 1	≥ 5.0	1.1	≥ 4.5	≥ 6.2
	mAb 2	≥ 4.4	1.8	≥ 5.1	≥ 4.8
	mAb 3	≥ 5.4	1.8	≥ 5.6	2.8
	mAb 4	6.4	<1.0		
	mAb 5	≥ 5.1	<1.0		
	mAb 6	≥ 5.9	<1.0		
	mAb 7	5.6	1.2		
	mAb 8	3.9	<1.0		
5.5	mAb 9	4.3	1.6		
	mAb 10	1.8	1.3		
6.0	mAb 11	1.3	1.8	5.2	2.6
	mAb 12	2.4	<1.0		
6.5	mAb 13	1.1	<1.0		

a. Monoclonal antibodies were purified on Fractogel SO_3- in bind and elute mode using various buffers and salts. Elutions were performed as either step or gradient elution's using increased levels of a sodium salt. Log reduction values were determined by $TCID_{50}$ assays.

Novartis *(M. Heitzmann)*

CEX chromatography was done with SP-Sepharose FF in the bind/elute mode. Data from a total of 13 validation studies with different antibodies (Table 26), showed that higher MuLV LRV obtained for product fraction correlated in the majority of cases with a lower overall virus infectivity spike recovery determined from all chromatography fractions (Flowthrough/pre-eluate, product fraction, post-eluate strip). In cases where a low overall spike recovery was obtained (e.g. < 1%) and associated with a high LRV in the product fraction (e.g. ≥4 log_{10}), this could be caused by inactivation of the sensitive MuLV in the chromatography system rather than by chromatographic separation, especially when the runs are performed at a pH of 5. Generally, in order to avoid the potential influence from low pH in combination with the shear stress by the chromatographic system on the virus, it might be better to use Q-PCR instead of infectivity assays for MuLV testing in a CEX step.

Table 26: Clearance and partitioning of MuLV (measured by $TCID_{50}$) in SP-Sepharose Chromatography. Total recovery of infectivity (Σ $TCID_{50}$) calculated from LRV's. Data courtesy of Markus Heitzmann, Novartis.

Antibody	Virus	LRV			Σ $TCID_{50}$
		Pre-eluate	Product eluate	Post-eluate/ strip	Total recovery
mAb 2	**X-MuLV**	>3.9	3.1	4.8	0.1%
mAb 3	**X-MuLV**	3.7	2.2	2.1	1%
mAb 4	**X-MuLV**	2.9	1.9	1.7	3%
mAb 5	**X-MuLV**	3.6	2.8	3.8	0.2%
mAb 6	**X-MuLV**	2.4	1.9	2.6	2%
mAb 7	**X-MuLV**	1.6	1.3	0.7	28%
mAb 8	**X-MuLV**	2	1.4	0.4	45%
study 1 mAb 4	**A-MuLV**	>3.6	5.5	3	0.1%
study 2 mAb 4	**A-MuLV**	>3.6	3.7	>3.3	<0.1 %
mAb 8	**A-MuLV**	>2.8	2.9	0.9	13%
mAb 9	**A-MuLV**	>3.5	3.9	1	10%
mAb 10	**A-MuLV**	>3.9	5.7	>4.4	<0.01 %
mAb 13	**A-MuLV**	>3.1	2.8	1.6	3%

Hydrophobic Interaction Chromatography (HIC)

Amgen *(L. Connell-Crowley)*

HIC has been used as a polishing step in some mAb manufacturing processes for removal of HMW and host cell impurities. Amgen data from several HIC operations using a variety of resins, salts, buffers, pH, and operating modes (BE or FT mode) indicate that some HIC steps can achieve effective X-MuLV clearance, although little or no MMV clearance is obtained. The effectiveness of HIC for X-MuLV removal correlates with the resin used (Fig. 26). The more hydrophobic resins such as Phenyl Sepharose High Sub and Butyl 650M appear to have the ability to achieve greater than 4 log_{10} of removal and, in some cases, have no detectable virus in the product pool. Understanding the mechanism and operating space for X-MuLV removal on HIC resins may some day allow for design of additional X-MuLV clearance into a process and provide information on the feasibility of generically validating X-MuLV removal by this step, although for now a product-by-product approach is taken.

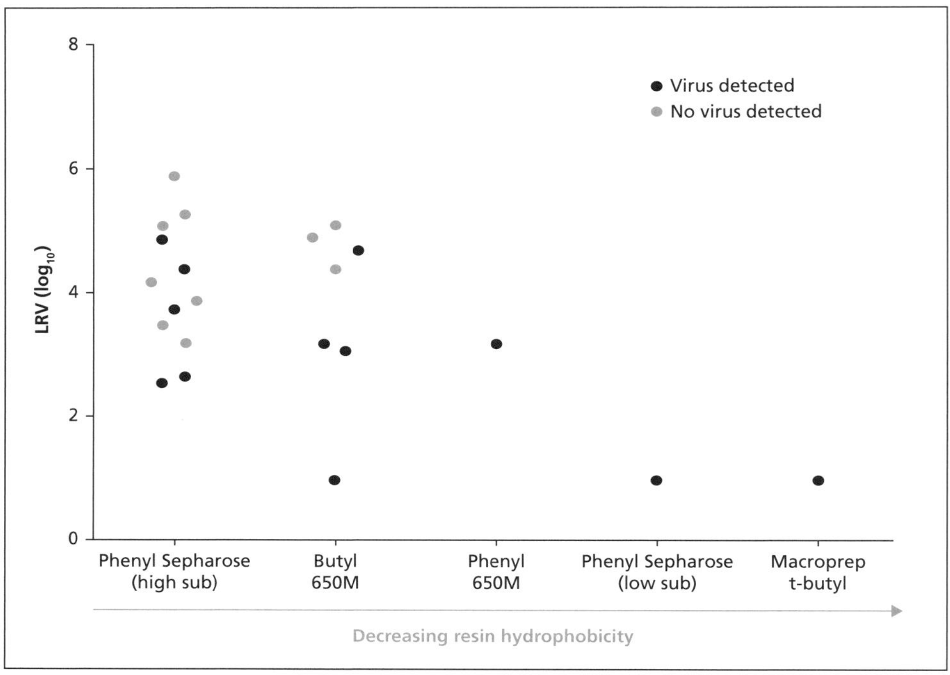

Fig. 26: Clearance (infectivity) of X-MuLV by HIC using undisclosed proprietary mAbs. HIC steps for different mAbs were operated in either the bind and elute or flow through mode using various buffers, salts, pHs and resins. All flow through steps were designed to achieve >90% yield, while some bind and elute steps had slightly lower yields. Log reduction values were determined by the $TCID_{50}$ assay. Data courtesy of Lisa Connell-Crowley, Amgen.

Chromatography with mixed-mode resins

Amgen *(L. Connell-Crowley)*

Virus retention was assessed using Captoadhere, a relatively new mixed-mode resin from GE Healthcare. It contains both anionic and hydrophobic functional groups. To evaluate the potential of Captoadhere for X-MuLV removal, X-MuLV binding and elution with a Captoadhere column was compared to that of Q Sepharose Fast Flow, a traditional anion exchange resin that is known to effectively remove X-MuLV [11, 19, 22]. X-MuLV was spiked into a low conductivity buffer at pH 6.0, 7.0 or 8.0, and loaded onto each column. Under all conditions, no virus was detected in the flow through fraction, indicating that the virus was bound to the column. The columns were then washed and eluted using a salt gradient. As shown in Figure 27, the virus was eluted from the Q Sepharose column at each pH, at greater than 250 mM salt. In contrast, no virus was detected in any elution fractions from the Captoadhere resin up to 800 mM salt, under the pH range tested. The data indicate that the Captoadhere resin may have a broader operating space for X-MuLV removal than Q Sepharose Fast Flow. Further definition of the operating space and robustness of both X-MuLV and MMV removal by Captoadhere could be used to facilitate generic validation of this step.

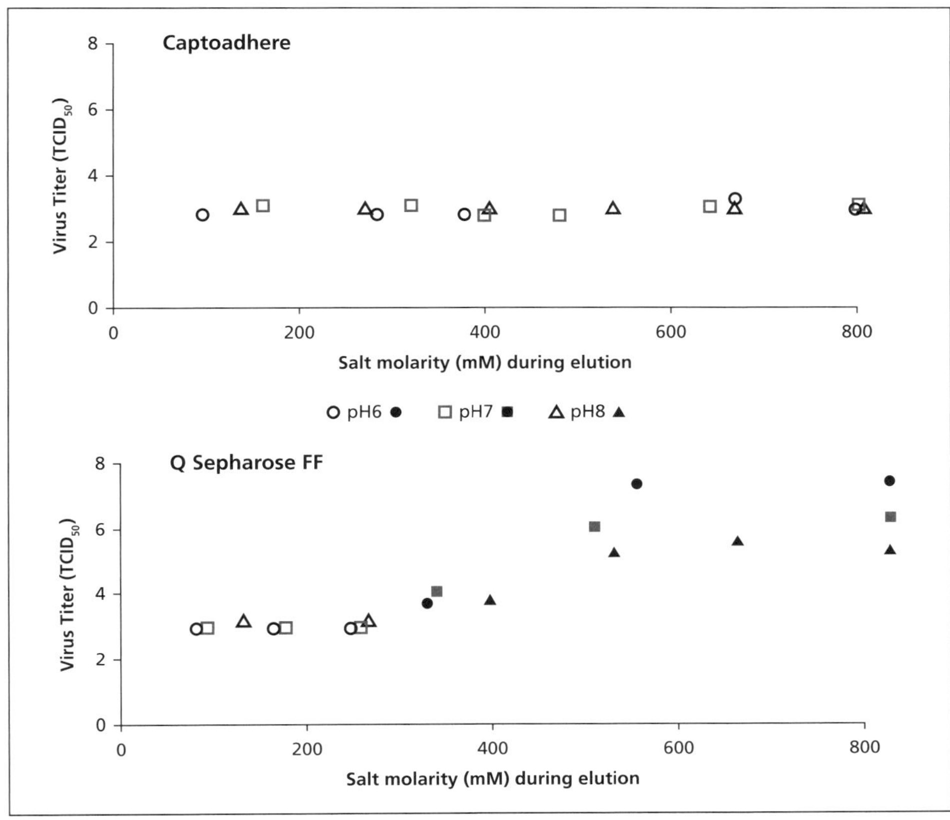

Fig. 27: Clearance (infectivity) of X-MuLV comparing Captoadhere and Q sepharose FF unit operations. A low conductivity buffer at pH 6.0, 7.0 or 8.0 was spiked with X-MuLV and loaded onto a Captoadhere or Q Sepharose FF column. The columns were washed and then eluted with a salt gradient. The flow through/wash fraction and fractions taken at various intervals during the gradient were assayed for X-MuLV by the $TCID_{50}$ assay. Open symbols indicate no virus was detected in that fraction. Data courtesy of Lisa Connell-Crowley, Amgen.

Centocor Ortho Biotech *(P. Alfonso and P. Alred)*

The influence of different load (g/L) and pH conditions on the removal of MMV and MuLV by CaptoAdhere were studied. The outcome confirms that CaptoAdhere removes viruses (MMV and MuLV) under a variety of processing conditions. The use of this resin adds significant value to the viral clearance capacity of the downstream process. The data are summarized in Table 27.

Table 27: Clearance of X-MuLV and MMV (LRV) by CaptoAdhere using a proprietary mAb (measured by infectivity). Data courtesy of Pedro Alfonso, Centocor Ortho Biotech.

Load (g/L)	pH	X-MuLV	MMV
20 – 25	8.5	≥4.8	4.8
100-150	6.7	≥5.5	≥4.7
64	5	≥5.0	ND

Column bed height ranged from 20 to 30 cm, flow rate was 300 cm/h for all studies.
ND= not determined

Overall Summary, Alternative Chromatographic Methods

CEX in BE mode is often used as a polishing step in mAb production and can be effective for removal of MuLV. It was shown that pH and ionic strength are important to achieve an elution profile that separates the mAb product from the virus. The operational range to achieve a reliable separation of virus and product is smaller than that observed for AEX and may change from one virus to the other. The effectiveness of virus removal by CEX is therefore virus specific. As was shown in the summary of the data from regulatory submissions as well as in the industry presentations, CEX may not be very effective for removal of a key adventitious virus, the murine parvovirus MMV.

HIC is especially used for removal of residual HCP and HMW protein, in addition to virus. Regarding the latter, data provided by Lisa Connell-Crowley (Amgen) suggests that the stronger the hydrophobic properties of the resin, the stronger the binding of MuLV to the resin and presumably the more efficient removal from the product stream. More work would be necessary, however, to understand the complex interaction between the resin, the product and the virus under the specific conditions of mAb production.

In contrast, the results with mixed-mode resins were very promising. Data with CaptoAdhere were presented. CaptoAdhere is a strong anion exchanger combined with hydrophobic interaction properties. It is comparable to other anion exchange chromatographic media, like Q Sepharose FF, but removal of MuLV seems to be more robust with respect to variations in load, pH and ionic strength. This resin was also effective for the removal of MMV (generally >4 $\log_{10}$). Mixed-mode resins like CaptoAdhere may add significant value to the virus clearance capacity of downstream processing, but further characterization is needed to understand the mechanism of action before general conclusions can be made about the value of this type of resin for viral clearance.

Miesegaes G, Bailey M, Willkommen H, Chen Q, Roush D, Blümel J, Brorson K (eds): Proceedings of the 2009 Viral Clearance Symposium. Dev Biol (Basel). Basel, Karger, 2010, vol 133, p 67.

SESSION VI

Virus Inactivation by Detergents

Miesegaes G, Bailey M, Willkommen H, Chen Q, Roush D, Blümel J, Brorson K (eds): Proceedings of the 2009 Viral Clearance Symposium. Dev Biol (Basel). Basel, Karger, 2010, vol 133, pp 69-74.

Virus Inactivation by Detergents

Technical Background

In this session, virus inactivation by classical SD treatment or by detergent-only treatments was discussed, whereas inactivation by low pH was covered separately (Session III).

S/D treatment using 0.3 or 1.0% tri-(n-butyl) phosphate (TNBP) together with 1% detergent, like Polysorbate 80 or Triton X-100, was originally developed for inactivation of blood products to remove the risk of contamination with HIV, HBV and HCV [27-30]. It is in general a very effective method for inactivation of enveloped viruses; the lipid-containing virus envelope is destroyed by reaction with these chemicals, resulting in a loss of the ability of the virus to attach to target cells. A similar effect can also be achieved even if detergents alone are used. The use of detergent alone at relatively low concentrations provides the added benefit that specific process steps for the removal of solvents are not needed (the detergent can be removed by unit operations which are in place for the purification of the mAb). Examples of both strategies were presented.

Regulatory Agency Presentation

FDA *(G. Miesegaes)*

We assessed the clearance power of three commonly used chemical inactivation methodologies: Low pH (described in Session III), S/D, and detergent-only (i.e., detergent inactivation in the absence of solvent). Detergent-only treatment is arguably more practical than S/D because solvents are known to adversely limit the signal to noise ratio of infectivity assays [31, 32], and they must be removed by downstream processing, thus adding undesirable time and resource requirements. In general, detergent-only processes are non-toxic, and test articles from validation studies can be assayed at lower dilution factors. Also, the imperative for removing detergents during commercial scale processing is less than that for S/D processes (failure to remove detergents is typically less consequential than failure to remove solvents). FDA presented viral clearance data for both S/D and detergent-only inactivation methods (Fig. 10).

Chemical inactivation dissolves or damages viral envelopes or envelope proteins; thus, non-enveloped viruses such as parvo- and picornaviruses (e.g. poliovirus) are not effectively inactivated using these techniques [3]. We therefore focused our analyses exclusively on MuLV, an enveloped retrovirus.

Previous publications reported that both low pH and S/D inactivation of retroviruses is robust [9, 27, 33, 34]. Similarly, our regulatory database found that low pH inactivation was quite effective in clearing MuLV (Session III). However, relative to low pH treatment (typically ranging from 4-7 LRV), ranges of clearance using detergent-based inactivation methods were generally lower (Fig. 10c-f) with the majority of entries claiming inactivation of <4 log_{10}. While records reporting complete inactivation may still be considered acceptable, the lower overall values may be a result of sample toxicity limitations (i.e., the samples required extensive dilution, thus lowering the LRV window). Not all entries cited complete clearance claims (7%) and therefore comprise a set of outliers which are currently under analysis. Of note, there were no claims of complete inactivation with values less than 2 log_{10} for either detergent-based inactivation method, whereas 11% of the records had reported >5 log_{10} of clearance.

Surprisingly, almost half of the S/D records did not claim complete clearance (incomplete: n=41; complete: n=44), even though this step is considered to be the regulatory standard for chemical inactivation [27, 33]. There are a number of plausible explanations for this, including (1) the result of reporting practices in early submissions, rather than poor performance by S/D; as the database included studies performed in the mid 1990's, reporting may not have paid sufficient attention to including ">" signs with the LRV numbers; and (2) experimental conditions used for titer determination were less than ideal; because of the high cytotoxicity of the S/D reagents, for instance, dilution of samples before testing is necessary and results in lower LRVs; (3) other factors such as temperature, sample mixing (i.e. continuous vs. static incubation) and virus preparation quality (i.e. crude or purified) may have influenced the ability to achieve complete clearance. Given these considerations, it is likely that many of the reported clearance values for S/D treatment are underestimates of the true process capabilities; claims that did not cite complete inactivation of MuLV for instance should be viewed as outliers, and not necessarily as a failure of the method.

We next isolated and analyzed detergent-only records (Fig. 10e-f), though it is difficult to make sweeping conclusions given the low numbers obtained (total n=24). We found, however, that similar to S/D treatment, LRV ranges cited in detergent-only records varied depending on whether complete inactivation was claimed. There were a substantial number of failed incomplete clearance records (defined here by LRV <2) (33%; Fig. 10e) and therefore make up another group of outliers. It is possible that these low-LRV outliers are due to the absence of a solvent-induced inactivation component. Alternatively, they could reflect differences in detergent type (e.g. Tween vs. Triton) or detergent concentration. Since solvents are not present, buffer toxicity limitations are more applicable to S/D inactivation studies than to detergent-only studies. Thus, low LRV detergent-only data are less likely to be underestimates of the process capabilities than in the case of the S/D outliers as described above. We did note however, that the majority of detergent-only records citing complete clearance achieved >4 log_{10} clearance of MuLV (Fig. 10f), again probably because detergent-only inactivation affords a significant experimental window. In conclusion, detergent-only inactivation appears to be robust, when and if properly designed.

Industry Presentations

Centocor Ortho Biotech *(P. Alfonso and P. Alred)*

S/D treatment using 0.2% TNBP and 1% Tween 80, was investigated. Under these conditions, the effect of different protein concentrations used in different products was analyzed on inactivation of X-MuLV. Table 28 provides a summary of the results.

Table 28: Clearance of X-MuLV (LRV) by solvent/detergent inactivation as a function of protein concentration (mg/mL) using six proprietary mAbs. Data courtesy of Pedro Alfonso and Patricia Alred, Centocor Ortho Biotech.

Product	Clinical studies, phase	Protein concentration (mg/ml)	X-MuLV
A	I	20.0	3.5
	II	20.0, 30.0, 40.0	≥3.6, ≥4.2≥4.1
	III	10.0, 40.0, 60.0	≥2.2
B	I	13.0, 17.0, 20.0, 24.0	≥3.1, ≥3.2, ≥3.3, ≥3.1
	III	13.0, 17.0, 24.0	≥3.1
C	I	21.0	≥2.5
	II	19.0, 63.6	≥2.6, ≥2.9
D	I	14.0, 28.0	≥2.6, ≥2.9
E	I	16.0, 26.0	≥3.7
F	I	14.0	≥3.7
	III	10,0, 30.0, 45.0, 60.0	≥2.9, ≥3.0, ≥2.9, ≥3.1

The data demonstrate that the protein concentration used in the range of 13 mg/ml to 60 mg/ml did not influence virus inactivation when other parameters like concentration of the chemicals, temperature and time of treatment are in an acceptable range. The variation in the reduction factors observed in the individual experiments related to the initial virus load and to the dilution of the samples before virus titration which were needed to reduce the toxic effect of the chemicals on the cells used in the virus assay. They do not reflect a difference in the kinetic of virus inactivation or the effectiveness of the method.

In addition to this first investigation, the influence of low variation in S/D content and the effect of temperature and protein concentration were studied in another experiment using ERV, another murine retrovirus (Table 29).

Table 29: Clearance of ERV by solvent/detergent inactivation as a function of protein concentration, time, temperature, and solvent/detergent ration using proprietary mAb "X". Data courtesy of Pedro Alfonso and Patricia Alred, Centocor Ortho Biotech.

Parameter	Value					
Ratio S/D stock to product	0.08	0.08	0.08	0.11	0.12	0.12
Temperature (°C)	21	15	25	21	15	15
Protein (g/L)	22.9	15	15	22.9	5	58.1
Inactivation time (minutes)	**LRV**					
2		0.9	0.9		0.9	0.5
10	≥3.4			≥3.4		
15		1.4	2.4		1	1.5
30	≥3.4			≥3.4		
45		3.4	≥3.4		3.4	≥3.0
60	≥3.4	≥3.5	≥3.4	≥3.4	≥3.5	≥3.0
90		≥3.5	≥3.4		≥3.5	≥3.0
120		≥3.5	≥3.4		≥3.5	≥3.0
360	≥3.4			≥3.4		

The greatest effects on virus inactivation kinetics were low temperature and short incubation time, which were interdependent. The trend in the influence of the protein concentration was not clear from the data and no effect of S/D ratio to product in the range studied was observed.

Genentech, Inc. *(Q. Chen and B. Yang)*

The effect of Triton X-100 on inactivation of MuLV was studied. In the first set of experiments X-MuLV inactivation by Triton X-100 at different concentrations and temperatures was compared with 1% Tween 80 and the classical S/D treatment (1% Tween 80 and 0.3% TNBP). In contrast to 1% Tween 80 where about 60 minutes of incubation time was needed to achieve an effective inactivation, 0.1% Triton X-100 inactivated X-MuLV to the limit of detection after ≤5 minutes. In this experiment, the treatment with 0.1% Triton X-100 was as effective as the classical S/D treatment. The results are summarized in Table 30 and Figure 28.

Table 30: Clearance of X-MuLV (LRV) by solvent/detergent and detergent only inactivation using a proprietary mAb process intermediate. Data courtesy of Qi Chen and Bin Yang, Genentech, Inc.

Condition	Inactivation time	X-MuLV
0.1% Triton X 100 @ 15°C	≤ 5 min	≥ 5.7
0.1% Triton X 100 @ R/T	≤ 5 min	≥ 4.7
0.5% Triton X 100	≤ 5 min	≥ 4.4
1% Tween 80	60 min	≥ 4.5
1% Tween 80 + 0.3% TNBP	≤ 5 min	≥ 4.8
0.1% Triton X 100 @ 15°C	≤ 5 min	≥ 5.7

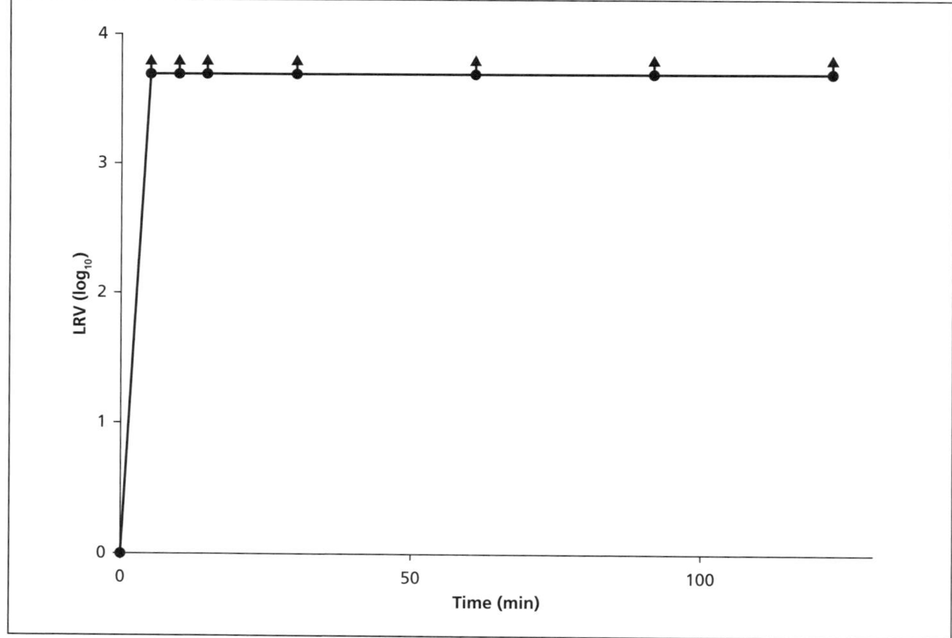

Fig. 28: Clearance of X-MuLV by detergent (0.1% Triton-X 100) inactivation. Data courtesy of Qi Chen and Bin Yang, Genentech, Inc.

In the second set of experiments several Triton X-100 concentrations were used for inactivation of MuLV in the mAb containing HCCF. The data are summarized in Table 31.

Table 31: Clearance of X-MuLV (LRV) by detergent inactivation using four proprietary mAb host cell culture fluids. Data courtesy of Qi Chen and Bin Yang, Genentech, Inc.

Triton concentration (%)	mAb 1	mAb 2	mAb 3	mAb 4
10 min, RT, Regular titration				
0.5		≥2.1	≥3.7	≥2.9
0.2	≥3.9	≥2.5	≥4.1	≥3.3
0.1	≥4.2			≥3.6
0.05	≥4.2			
0.02	≥4.4			
0.01	0.2			
0.005	0.5			
60 min, RT, Large volume testing				
0.2	≥5.9	≥5.9	≥5.7	≥6.0

A Triton X-100 concentration of 0.2% was effective for inactivating X-MuLV in the HCCF of four antibody products and also residual infectivity was not detected when a larger sample volume was tested at the end of the incubation time (60 minute time point, at room temperature).

Data from both experiments demonstrate that the detergent (Triton X-100) is effective for inactivation of enveloped viruses (e.g. MuLV) and provides an orthogonal step for retroviral clearance that might be amenable to modular validation.

Overall Summary, Virus Inactivation by Detergents

The classical S/D method using 1% or 0.3% tri-(n-butyl) phosphate (TNBP) and 1% detergents (e.g. Tween 80, Triton X-100, etc) over 4 to 6 hours is very effective in inactivating enveloped viruses. This method was developed and successfully used for manufacture of safe and efficacious medicinal products derived from human plasma such as coagulation factors and immunoglobulins [28-30]. Neither product class, protein concentration, process temperature, nor solution pH have a significant impact on virus inactivation when constrained within ranges typical of manufacturing [27]. The robustness and high effectiveness of the S/D methods has now been shown for mAb products. With regard to inactivation kinetics, temperature and time had the greatest impact. All parameters tested resulted in a complete inactivation of the retrovirus used in these experiments. Despite the high effectiveness of S/D treatment, low LRVs were reported, most likely because samples must be diluted significantly before testing (1:50 to 1:100 v/v) to reduce cytotoxicity and to stop inactivation.

Another drawback of S/D inactivation is the need to remove the chemicals (i.e. solvents) during commercial manufacture, which is typically achieved with specific chromatographic steps. It is therefore very promising that detergent alone affords significant LRVs. The effectiveness of detergent-only inactivation eliminates the need to remove TNBP. The use of detergent-only inactivation in the cell culture harvest may also streamline the process, as removal of the detergent can be achieved at the capture step and eliminate the need for specific chromatography steps. The viral clearance database provided in this session supports the assumption that Triton X-100 at relatively low concentrations may provide a robust and reliable method for inactivation of MuLV and other enveloped viruses. It was shown that MuLV was completely inactivated when cell culture harvest was treated with 0.2% Triton X-100 over 60 minutes at room temperature. However, it is not clear whether the variable composition of cell culture harvests caused by different levels of cell lysis, different buffer composition, and lipid content, may require higher detergent concentrations to assure virus inactivation. Further studies are needed to substantiate the robustness and reliability of this method.

Miesegaes G, Bailey M, Willkommen H, Chen Q, Roush D, Blümel J, Brorson K (eds): Proceedings of the 2009 Viral Clearance Symposium. Dev Biol (Basel). Basel, Karger, 2010, vol 133, p 75.

SESSION VII

Virus Filtration

Miesegaes G, Bailey M, Willkommen H, Chen Q, Roush D, Blümel J, Brorson K (eds): Proceedings of the 2009 Viral Clearance Symposium. Dev Biol (Basel). Basel, Karger, 2010, vol 133, pp 77-91.

Virus Filtration

Technical Background

Virus retentive filtration (often incorrectly referred to as "nanofiltration") is an industry standard method for removing viruses and is employed in almost all modern bioprocesses. Two main classes of filters exist: large virus (e.g. retroviruses) and small virus (e.g. parvovirus) retentive filters [35, 36]. In general, newer processes use dead end (direct flow) mode virus retentive filters, while older process trains use tangential flow filtration. Compared to other methods (e.g. low pH and detergent inactivation), virus retentive filtration is less likely to damage the product. Furthermore, LRVs are more predictable when this unit operation is well designed and known pitfalls are avoided (e.g. virus passage under filter overloading conditions; [37]). Important process parameters commonly associated with virus retentive filters include transmembrane pressure, filter pore size (i.e. large vs. small virus), flux rate per filter surface area, overall volumetric throughput and process fluid flux as a percentage of initial buffer flux (for some filter brands) [38]

Regulatory Agency Presentations

FDA *(G. Miesegaes)*

FDA grouped filtration records into two categories (Fig. 29), large virus retentive filters (Pall DV50, Millipore NFR, Asahi Kasei Planova 35N) and small virus retentive filters (Pall DV20, Millipore Viresolve NFP, Vpro, V70 and V180, Asahi Kasei Planova 20N and 15N, Sartorius Virosart CPV). Vendor-claimed retentive properties (i.e., small vs. large virus) and PDA rating nomenclatures [39] of a given filter were used to assign records into either of the two size categories. Brand names were excluded.

Analysis revealed that complete clearance claims were made for MuLV in the vast majority of records (88%, large virus filters; 95%, small virus filters; Fig. 29a,b). Complete clearance of MuLV (a 80-100 nm virus) by virus filters is expected, given that the same filters can clear >8-9 $\log_{10}$ of the smaller (64-82 nm; [40]) bacteriophage PR772. While it seems unusual that complete MuLV clearance was not claimed in all cases, this could represent similar flaws in reporting practices as described above (i.e. omission of a ">" LRV sign in the regulatory submission), or technical issues such as splash-overs during filtration, detection of naked DNA by PCR assays, etc. In addition, even for cases where complete clearance was not claimed, substantial removal (>5 $\log_{10}$) was achieved (data not shown). Clearance of MuLV using small virus filters was lower on average (e.g. 44% large filter records reported 5-6 $\log_{10}$ clearance, compared to 27% of small filter records). This may be due to the fact that small virus retentive filters are more susceptible to clogging/fouling, which constrains the experimental window by limiting the spike load and the subsequent LRV that can be measured in a validation study [36].

Although some studies using large virus filters existed (n=13), most of the MMV records reflected the use of small virus filters (Fig. 29c,d). As might be expected, the majority of MMV records using large virus retentive filters failed to report a substantial clearance value

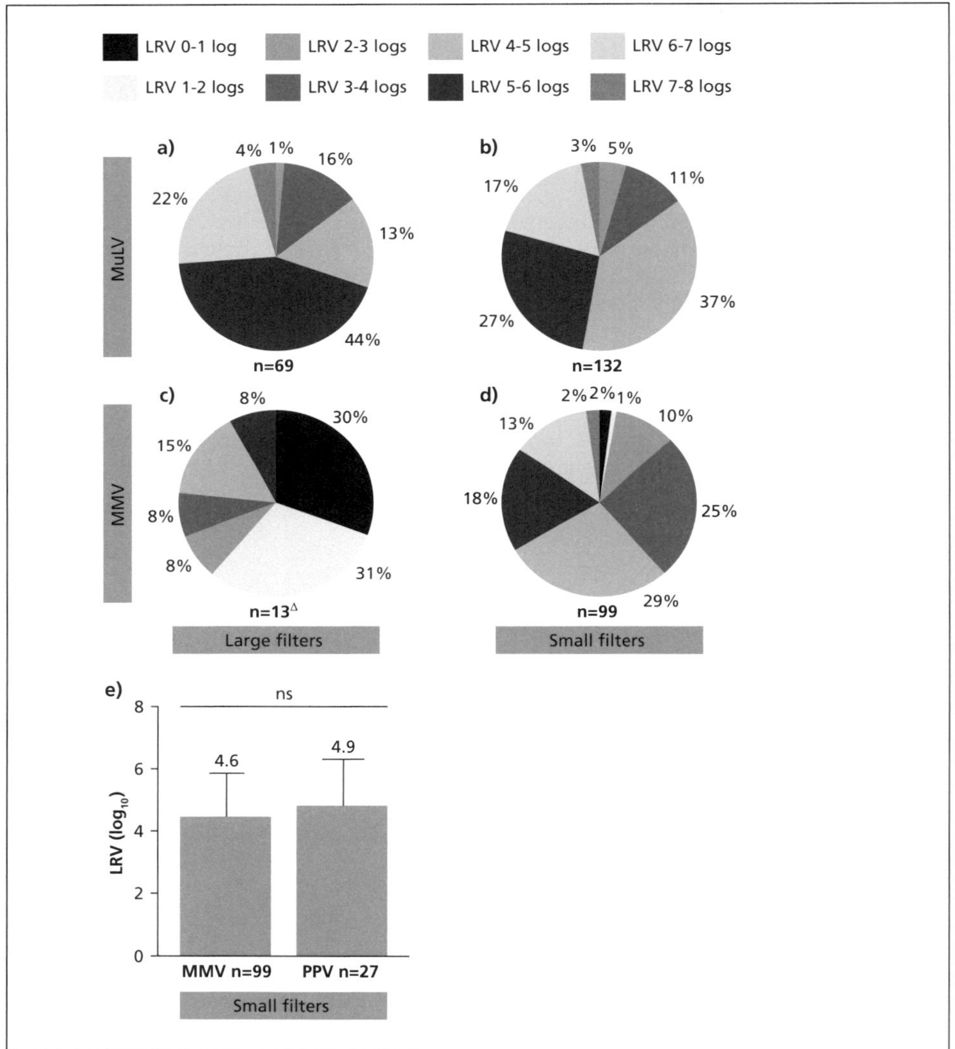

Fig. 29: Analysis of virus retentive filtration records in CDER's viral clearance database. LRV ranges from large and small virus retentive filtration records binned into 1 $\log_{10}$ intervals for (a,b) MuLV and (c,d) MMV. (e) Removal of MMV vs. PPV using small virus retentive filters. Bars are standard deviations. ΔLow number of records noted. ns = not significant. Bars are standard deviations. Data courtesy of George Miesegaes, FDA.

(Fig. 29c), with 60% reporting less than 2 log_{10}. Moreover, none of the records claimed complete removal (data not shown). Therefore, MMV removal is unlikely to be effective unless small virus filters are used. It should be noted that of the records claiming 2-6 log_{10} retention of MMV by large virus filters (comprising 39% of this group), this level of clearance is most likely due to use of crude virus preparations or a product matrix that induces virus aggregation [31]. Records reporting the use of small virus filters revealed that MMV was effectively removed (Fig. 29d); only 3% reported less than 2 log_{10} and of those reporting complete clearance, 84% claimed retention values above 4 log_{10}. Records that did not claim complete clearance (78%) largely claimed retention values above 3 log_{10}. Although 95% of records using small virus filters for MuLV demonstrated complete clearance, only half (51%) of the MMV records demonstrated the same.

The predominant mechanism of particle retention by liquid filters run under typical, non-overloading conditions is size-based sieving [36, 41]. Therefore, we investigated whether clearance studies performed on a second parvovirus of similar size would yield identical values. We compared filter records of MMV and PPV and found that the mean clearance values are indeed quite similar (p=0.2; Fig. 29e).

As noted above, a small number of MuLV filtration records did not claim complete clearance. We do not believe that this reflects a failure of filtration technology but rather the limitations of retrospective meta-analyses from our regulatory data sources. As noted earlier, our dataset consists of records from a broad range of firms, some of which are highly experienced. In addition, some of the records are 20+ years old, and were submitted before viral safety studies and reporting formats were standardized. Thus, for some records it is not immediately apparent whether, for example, the study was run properly (i.e., whether aerosols or inappropriate contact with liquids were avoided, filter integrity testing was performed, steps to eliminate false signals from free nucleic acids were taken if PCR assays were used, etc.); or if recordkeeping errors or omissions had been introduced (i.e., complete clearance was observed but the IND sponsor failed to include a ">" symbol when typing up the report for submission to FDA). In the absence of additional information, any conclusions regarding the inadequacy of filtration technology are unwarranted.

Overall, the FDA regulatory database revealed that virus retentive filtration is robust. Large virus retentive filters generally remove 5+ log_{10} of MuLV, and small virus retentive filters typically remove 3+ log_{10} of parvovirus. Often, the sensitivity of the assay or spiking limitations are the limiting factor for the LRV that can be measured.

PEI *(J. Blümel)*

Most antibodies (74 of 91) were processed through filters designed for reduction of small viruses. 15 antibodies were processed through filters designed for removal of large viruses such as retroviruses. Only 2 production processes did not include a virus filtration step. PEI observes a trend towards use of filters designed for small virus retention. Removal of retroviruses was effective and robust with all types of filters and with all different antibodies. A breakthrough of infectious retroviral particles was never observed in cases where filtration was validated using infectivity assay.

Reduction of small non-enveloped viruses such as parvoviruses or picornaviruses was more variable. Table 32 summarizes the data from small virus filtration. In most cases (n=59), parvovirus reduction factors were clearly on the order of 4 (3.5-4.5) log_{10} or above. However, moderate reduction factors (<3.5 log_{10}) with break-through of infectious virus were observed in 10 of 69 cases (14%). Six of 69 cases (9%) had reduction factors lower than 3.0 log_{10}. There was no obvious mismanagement of filters (e.g. overloading, clogging of filters) in all 10 study reports and no other factor influencing the reduction data could be identified. MMV was used in 46 studies and PPV was used in 23 studies. There was no correlation between the species of parvovirus (MMV versus PPV) and the magnitude of reduction factor. Information about the method of virus spike preparation was not systematically described in the study reports.

Table 32: Number of antibodies grouped to small virus retention filters and reduction factors for MVM or PPV. Data courtesy of Johannes Blümel, PEI.

Log_{10}-RF	Filter A	Filter B	Filter C	Filter D	Filter E	Filter F[d]
>4.5	11[a](5)[b]	5[a] (3)	14[a] (6)[b]	10[a] (3)[b]	1[a] (1)[b]	1 (1)
3.5-4.5	3[a] (1)[b]	5[a] (0)[b]	6[a] (2)[b]	1[a] (0)[b]	1[a] (0)[b]	0
<3.5	3[a] (0)[b]	2[a] (1)[b]	4[a] (0)[b]	2[a] (0)[b]	0	0
No data	0	5[c]	0	0	0	0

[a] total number of antibodies

[b] Number of antibodies where no residual virus was detected in the filtrate is indicated in brackets

[c] Polio virus was used instead of a parvovirus.

[d] not a filter designed for virus retention

In summary, clearance of large viruses such as retroviruses was consistent using either filters designed for retention of large viruses or small viruses. Effective reduction of parvoviruses could be demonstrated in most cases using small virus retention filters. However, breakthrough of parvoviruses occurs sometimes and it seems prudent to carefully validate reduction with respect to small viruses such as parvoviruses.

Industry Presentations

Overview

The industry presentations were from both medium- and large-sized firms, including some DoE data from the larger firms. While the experience of individual firms varied to some extent, general trends were observed. The industry data from the virus filtration session focused primarily on more challenging filtration applications, namely the removal of small viruses (e.g. parvovirus, poliovirus) by small virus retentive filters, (also known colloquially as "parvofilters"). Data was presented using filters from all four firms that manufacture virus filters: Millipore Viresolve NFP, Millipore VPro, Pall Ultipor DV20, Asahi Kasei Planova 15N & 20N, and Sartorius Virosart CPV filters. All of these filter types are designed to target removal of smaller (e.g. parvo or polio) viruses. The industry presentations, particularly those focusing on DoE data, addressed the impact of key process parameters (pH, ionic strength and protein concentration), operating parameters (flow rate, transmembrane pressure, volume and mass load, flux) and equipment parameters (filter type) on the performance of virus retentive filtration. Flux decay was also evaluated, particularly for filter types where this parameter has been previously noted to be important (i.e. Viresolve NFP; [37, 38]).

Effect of viral preparation quality on virus filter throughput

Appropriate design of scale-down models and spike quality of model viruses was the subject of a number of presentations. This topic generated much discussion as groups from Amgen, Genentech, and Wyeth provided data that showed that spiking itself impacts the flux and filter capacity in ways that may not be representative of the full-scale operation. Similar observations from multiple groups provide an imperative for improvements in approaches for virus spike quality, as proposed by PDA Technical Report 47 from the Virus Spike Preparation Task Force [31].

Wyeth *(G. Bolton)*

Wyeth presented data demonstrating that virus spiking (0.25-1.0% A-MuLV) can accelerate filter fouling of both Viresolve Pro (dilute mAb spiked at 0.25%) and Virosart CPV filters (buffer spiked at 1% A-MuLV, Fig. 30). The fouling rates of these two filters should not be compared because they were run with two different process fluids as mentioned above.

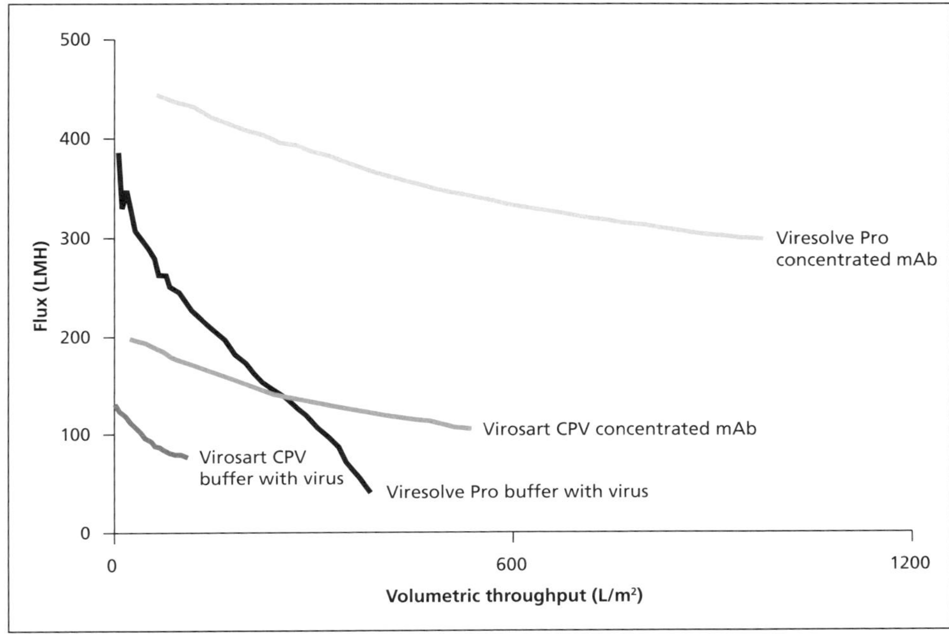

Fig. 30: Virus spiking (murine retroviruses) can accelerate filter fouling of both Viresolve Pro and Virosart CPV filter brands. Data courtesy of Glen Bolton, Wyeth.

Amgen *(E. Gefroh, V. Raol and H. Dehghani)*

One of the challenges with the use of commercially available parvovirus removal filters in mAb production processes is the capacity reduction often experienced during virus spiking studies. This reduction limits the allowable throughput for commercial manufacturing, ultimately increasing the cost for a step that accounts for about one-fifth of the entire downstream direct cost-of-goods. The drop in capacity, in some applications, can be attributed to decoupling an adsorptive pre-filter from the virus filter prior to spiking, which is required in order to demonstrate size-based removal across the virus filter. In addition, the virus spike itself often has a negative effect on the filtration performance [42, 43]. However, the virus preparations used during validation are not routinely characterized, and therefore the impact of virus preparation quality on filter performance is not well documented. The Amgen Process and Product Development group optimized the virus production process to improve the quality of the virus preps, as well as to study the effect of virus preparation quality on virus filter throughput. The results are summarized below.

The protein content of the virus preparations was used to guide the optimization of the virus production process with respect to the quality of the virus preps. The average protein content of the in-house X-MuLV preparations generated using a base process [44], and formulated in a BSA-containing buffer was approximately 5500 µg/ml. First, BSA was removed from the storage media to eliminate the unnecessary re-introduction of a protein impurity. No impact to virus stability was observed with this change [44]. A tangential flow filtration (TFF) based concentration step was then introduced prior to the ultracentrifugation (UC) step to increase overall production efficiency, although it resulted in an increase in the impurity content of the final prep. Additionally, the serum level (FBS) in the cell culture media was subsequently reduced from 10% to 2%. This change led to a significant decrease in the protein impurity levels, with no impact on virus titer. Finally, a buffer wash of the virus pellet followed by a UC step was included after the TFF-UC step to further reduce the impurity levels and re-concentrate the virus. These changes collectively decreased the protein content in the virus preparations from 5500 µg/ml to 50 µg/ml. The step-wise improvements in the virus production process are summarized in Table 33. The protein content of the in-house MMV preparations is also shown in the table, indicating that the base MMV process [44] generates preparations with relatively low protein impurity levels.

Table 33: Impact of modifications in the virus production process on virus filter performance (0.3% virus spike in pH 5 buffer, NFP filter, clearance measured by infectivity). Data courtesy of Eva Gefroh, Viveka Raol, and Houman Dehghani, Amgen.

Virus	% FBS in media	Harvest process	Storage media (+/-BSA)	Avg titer ($\log_{10}TCID_{50}$/mL)	Avg protein impurity (µg/ml)
X-MuLV	10%	UC	+	7.53	5513
	10%	UC	-	7.53	501
	10%	TFF	-	7.67	1715
	2%	TFF	-	7.55	432
	2%	TFF + wash	-	7.49	51
MMV	5%	UC	-	7.73	<32

The effect of X-MuLV preparation quality on virus filter throughput was then tested by spiking 0.3% of each stock into a pH 5 buffer, and assessing the flux profile on an NFP filter. The results from these experiments are shown in Figure 31. It is clear from these results that the step-wise improvements that led to a decrease in protein impurity in the preparations also correspond to an improvement in filterability. In fact, the cleanest X-MuLV preparation resulted in a flux profile similar to that of an MMV prep. The achievable throughputs during virus clearance studies are dependent on the quality of the virus prep, and particular attention must therefore be paid to the characterization of the virus preparations prior to validation studies.

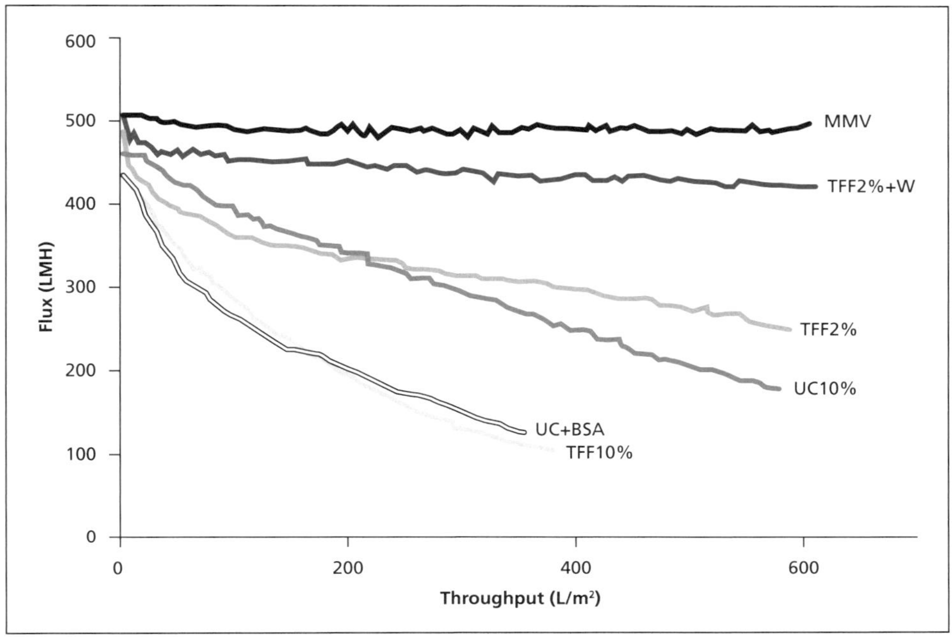

Fig. 31: Impact of modifications in the virus production process on virus filter performance (0.3% MMV virus spike in pH 5 buffer, NFP filter, clearance measured by infectivity). Letters correspond with descriptions in Table 35. Data courtesy of Eva Gefroh, Viveka Raol, and Houman Dehghani, Amgen.

Large virus removal

Information presented during the workshop using small virus retentive filters (>100 studies) confirmed the capability of these filters to effectively remove large viruses such as retroviruses (Table 34). Based on this work, in the future industry may be able to take a modular approach to filtration of retrovirus when small virus retentive filters are used. Given the fact that large virus retentive filters have been rated to remove 6 $\log_{10}$ of a 64-82 nm bacteriophage [36], it is not unreasonable to assume that small virus filters could function at least equivalently for large viruses.

Table 34: Overall numbers of studies of retrovirus across small virus filters as discussed during the virus filtration workshop.

Company	Number of antibodies tested
Human Genome Sciences	8
Novartis	13
Pfizer	18
Wyeth	5
Amgen	25
Lilly	12
Genentech	14
Boehringer-Ingelheim	14

Genentech: Proposal of an alternative method to measure retrovirus clearance across parvovirus sized filters *(Q. Chen and R. Specht)*

Filtration with Viresolve Pro can exhibit a reduced volumetric throughput capacity due to fouling by the mammalian virus spike (X-MuLV); this leads to a small-scale study that is not representative of manufacturing scale in terms of fouling as measured by flux decay (Fig. 32). In contrast, lowering the level of spiking should diminish the impact on permeability. Genentech described an alternative virus validation strategy using bacteriophage to overcome the current issues with mammalian virus stock solutions. Bacteriophages phiX-174 and PR772 have both been evaluated as alternative model viruses for parvovirus filtration. The advantage of the bacteriophage model viruses is that they can be produced to have very high titers and high purity and thus do not reduce/impair filtration capacity. Virus removal capacity of mammalian and bacteriophage viruses are comparable as shown in Table 35.

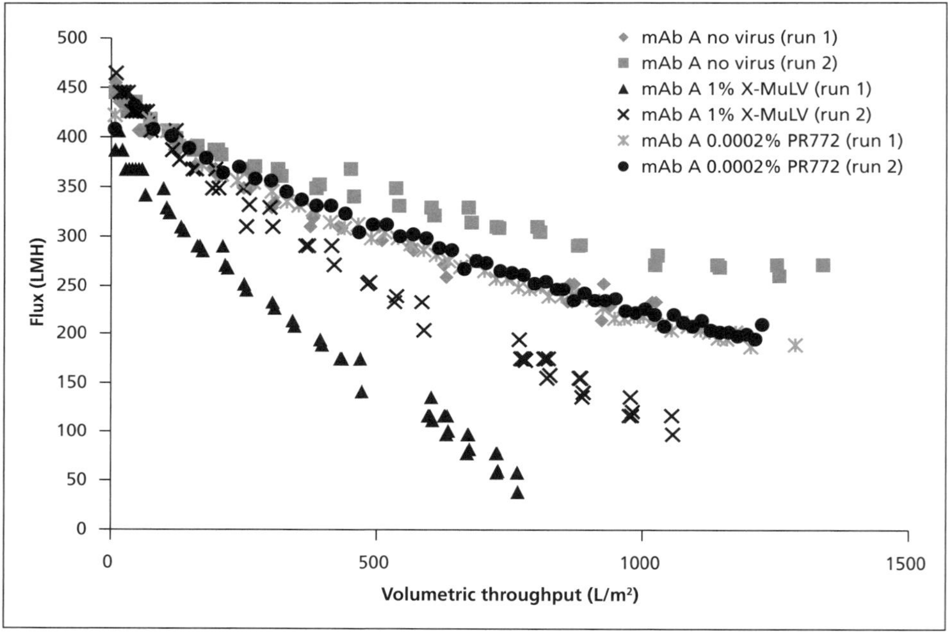

Fig. 32: Comparison of volume throughput for mAb A with no virus spike, 1% X-MuLV spike and 0.0002% bacteriophage PR772 spike. Data courtesy of Qi Chen and Rachel Specht, Genentech, Inc.

Table 35: Clearance of X-MuLV, PR772, MMV, and PhiX-174 by small virus retentive filtration using undisclosed proprietary mAbs. Data courtesy of Qi Chen and Rachel Specht, Genentech, Inc.

Model virus	Model virus size (nm)	Log reduction value
X-MuLV	80-110	≥ 4.8
PR772	64-82	≥ 5.3
MMV	18-24	≥ 4.2
PhiX-174	26-28	≥ 5.2

Flow decay and virus breakthrough

Several presentations dealt with the relationship between flow decay and virus breakthrough. This phenomenon has been observed previously with bacteriophage and MMV and was confirmed by the data presented by several firms.

Pfizer *(P. Mensah and M. Gustafson)*

Using 4 filters (Planova 20N, DV50, DV20, and Virosart CPV) in 26 studies with 18 mAbs challenged with either X-MuLV or MMV, Pfizer showed robust clearance of X-MuLV on all four filters (Table 36). However, difference in the flux profiles was observed among the filters (data not shown). Additional data presented suggest that the newer filters such as BioEx and Virosolve Pro may provide higher flux profiles without compromising LRV.

Table 36: Clearance of X-MuLV and MMV by small virus retentive filtration using undisclosed proprietary mAbs (measured by infectivity). Data courtesy of Paul Mensah, Pfizer.

Nanofilter	**# of mAbs**	**Average LRV**	
		X-MuLV	MMV
DV50	3	>5.33 ± 1.09	NA
DV20	11	>5.56 ± 0.94	>4.31 ± 1.45
Virosart CPV	1	>6.92	>5.82
Planova 20N	11	>5.24 ± 0.69	>5.16 ± 1.34

Human Genome Sciences *(O. Galperina)*

The effect of flow decay on viral clearance was evaluated by HGS using both large (X-MuLV, Reo-3 and PRV) and small (MMV) virus. They used two different Mabs and employed flow decay as an indicator for virus retention by a Millipore NFP filter when used in conjunction with a commercial charged depth pre-filter. The results showed no impact on clearance as a result of flow decay for large virus, but a significant effect was observed for small virus where breakthrough occurred at approximately 75% flow (Fig. 33). Based on their small scale data, HGS proposed that in production the flow decay be limited to less than 75% to provide assurance that the filter remains effective for virus retention.

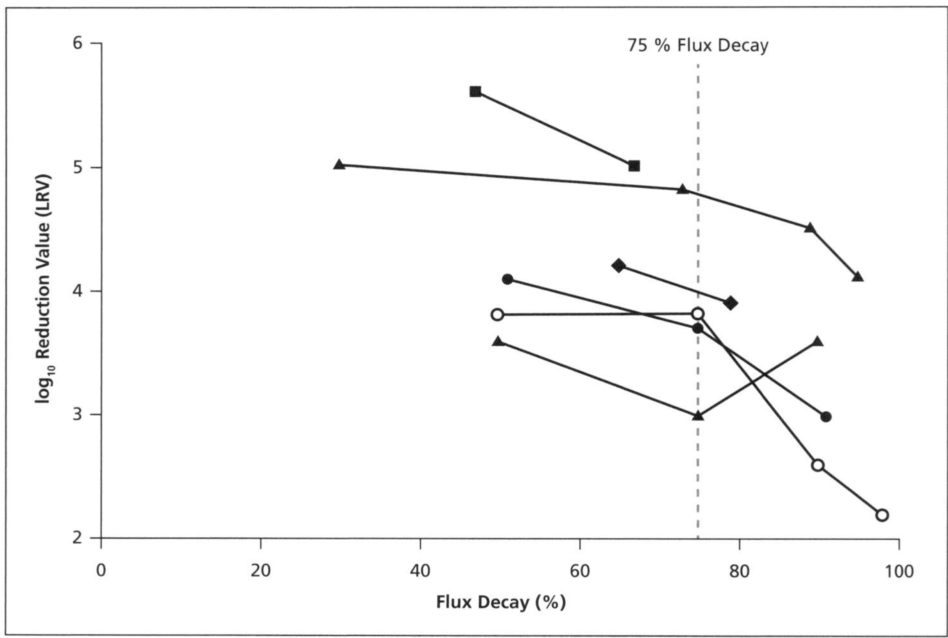

Fig. 33: Clearance of MMV by small virus retentive filtration as a function of flow decay, using undisclosed proprietary mAbs (measured by Q-PCR). The majority of points at low flow decay are "complete clearance" (i.e. no virus in filtrate) and at high flow decay absolute clearance values are measured (i.e. measurable virus in filtrate allows calculation of the precise clearance power). Symbols represent individual experiments. Data courtesy of Olga Galperina, Human Genome Sciences (HGS).

Centocor Ortho Biotech *(P. Alfonso)*

The correlation between flux decay and virus breakthough was confirmed by data presented by Centocor Ortho Biotech. Due to the reduction in flux caused by introduction of the virus spike, the full mass/volumetric load could not be achieved in the presence of the spike. Therefore, a study was conducted using two approaches:

- Spike-run: Virus was spiked into the product and the process step conducted with fractions taken at defined flux decay points.
- Run-Spike: Product was run across the filter to defined flux decay points, the process paused, spike introduced, then the run continued with fractions taken at defined flux decay points.

As can be seen in Figure 34a, virus breakthrough can be observed independent of volumetric/mass loading. However, Figure 34b shows that when breakthrough is observed, it is only seen at high flux decay regardless of the spike-run or run-spike design. This is consistent with published literature [38, 45].

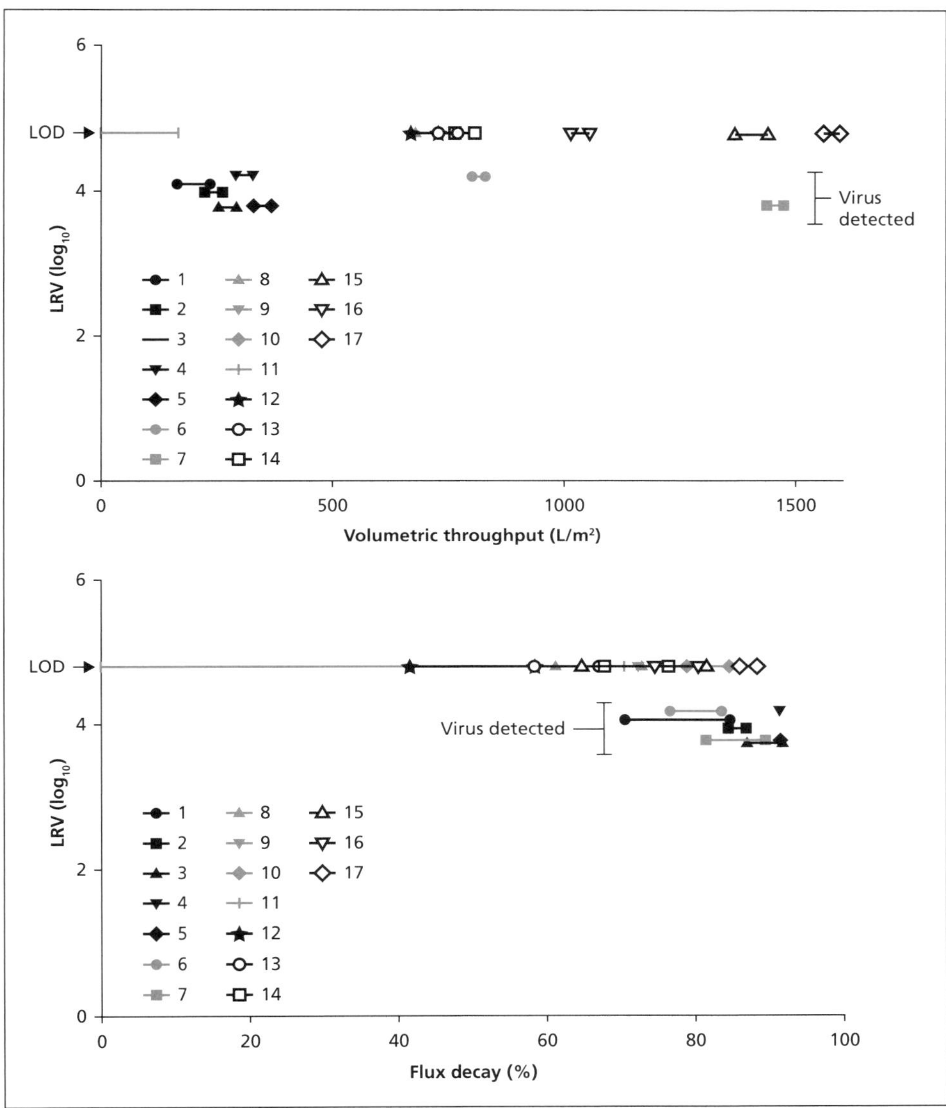

Fig. 34: Clearance of a small model virus by small virus retentive filtration as a function of flux decay (measured by infectivity) in a compilation of 17 filter runs. (a) LRV plotted as a function of mass loading. (b) LRV plotted as a function of flux decay. Data courtesy of Pedro Alfonso, Centocor Ortho Biotech.

Thus, the Centocor Ortho Biotech data verified that flux decay is the appropriate end point for filtration by certain small virus filters and is a process parameter that can be monitored and controlled to ensure processing within validated ranges.

Other observations

Novartis: Optimizing spike dilution *(M. Heitzmann)*

Filter load, throughput, capacity, cost and validation were discussed by Novartis. They recommended that validation of large throughputs should be performed at low virus spike to avoid virus filter clogging caused by virus overload. For example, in the case of Virosolve NFP they noted that parvovirus must be spiked (depending on the process solution) at less than 0.1% to reach the necessary process volumes during validation studies (Table 37).

Table 37: Virus spike (MMV), flow performance and LRV in virus removal studies on Virosolve NFP. Data courtesy of Markus Heitzmann, Novartis.

Antibody	pH	Spike MMV (Vol%)	Filter load (L/m^2)	Filtration time (min)	Flux mean (L/m^2/h)	Flow decay (%)	LRV
mAb 1	6.5	0.5	57	22	15	75%	4.6
mAb 2	5.5	0.5	151	35	259	15%	3.7
mAb 3	6	0.5	517	186	166	80%	> 4.6
mAb 4	6	0.03	248	80	186	40%	> 4.2
mAb 5	6	0.5	78	40	117	20%	> 4.3
mAb 6	6	0.25	209	50	250	25%	> 6.9
mAb 7	5.5	0.5	161	300	32	80%	6
mAb 8	6	0.25	145	30	290	5%	5.7
mAb 9	5.5	0.1	177	155	69	85%	> 6.3
mAb 12	5	0.05	166	133	75	95%	5.6
mAb 13	6.6	0.03	230	60	230	35%	3.8

Boehringer Ingelheim: Consistency of clearance *(F. Nothelfer)*

BI Pharma presented a study of the impact of volumetric capacity (50-350L/m^2), load (100-4000 gm/m^2), pH (5-8) and protein concentration (1.0 to 25mg/mL) on the performance of the Planova 20N virus retentive filter. With exception of a small dependence on pH, the Log Reduction (LRV) values were independent of all parameters studied. Enhanced clearance due to pH dependent virus aggregation has been reported to occur in some instances, although this is believed to be process fluid and virus spike quality dependent [31].

Eli Lilly and Company: Capacity variability *(M. Bailey and D. Chen)*

Data presented by Eli Lilly and Company showed that there is significant variability in capacity among individual laboratory scale filter devices for major brands of parvovirus filters. Five parvovirus filters from four vendors (Millipore NFP, Millipore Vpro, Pall DV20, Sartorius CPV, and Planova 20N) were evaluated using BSA as a surrogate protein. The results indicated that the overall capacity variation can be as great as ± 40%. In addition, using bacteriophage PP7 as a model virus, Eli Lilly and Company showed that the relationship between virus breakthrough and flux decay vary from brand-to-brand. Specifically, flux decay correlates with PP7 breakthrough for NFP, DV20, and Planova 20N with varying severity. On the other hand, flux decay does not lead to significant PP7 breakthrough for CPV and Vpro (Fig. 35).

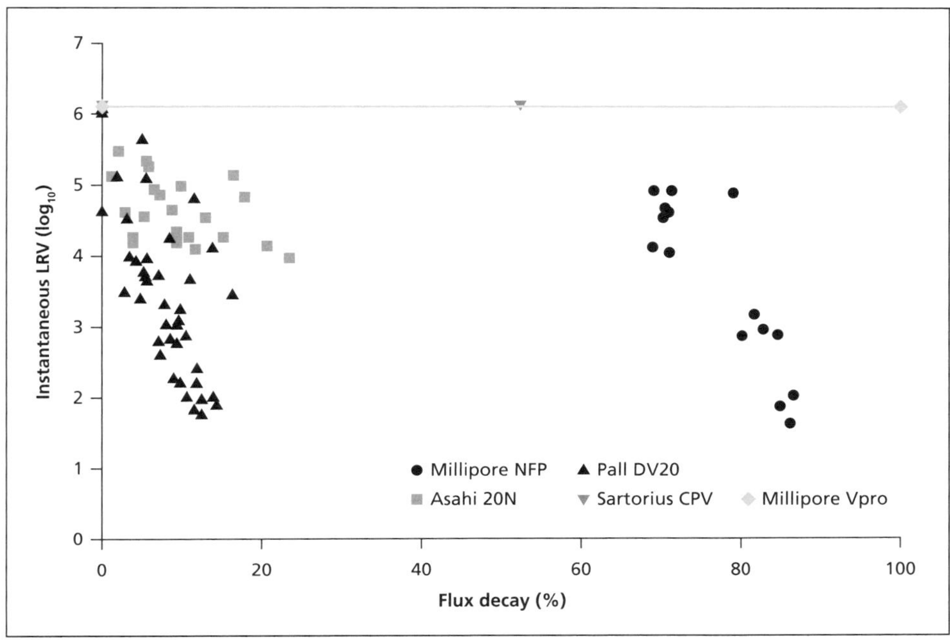

Fig. 35: Clearance of bacteriophage PP7 by small virus retentive filtration using undisclosed proprietary mAbs. Note that "Instantaneous LRV (log_{10})" corresponds to removal from instantaneous (i.e. grab) samples. Data courtesy of Mark Bailey and Dayue Chen, Eli Lilly and Company.

Wyeth: Proposal for filter reuse *(G. Bolton)*

Wyeth presented a proposal for virus retentive filter re-use. Small virus retentive filters make up a significant portion of the cost of mAb production; thus strategies to improve throughput and flux significantly impact process economics. The idea of reusing parvovirus filters after cleaning has been proposed to reduce costs and is based on a similar strategy used with many chromatography steps. To safely carry out membrane reuse in the manufacturing of biotherapeutics, a number of considerations would have to be taken into account. One argument in favor of this approach is that lower flux filters that do not typically plug extensively during operation could be cleaned with caustic and/or detergent to recover flux. Wyeth presented data from an Asahi Planova 20 device cleaned after use with biotherapeutic protein streams with or without the presence of virus spikes (Fig. 36). In theory, this filter could be reused, at least based on liquid flux. Additional challenges need to be addressed however, which include: (1) validating continued viral clearance under conditions of reuse; and (2) demonstrating absence of virus carry-over. While the participants appreciated the cost savings of such an approach there were concerns regarding regulatory acceptance, as well as validation and implementation in a manufacturing setting.

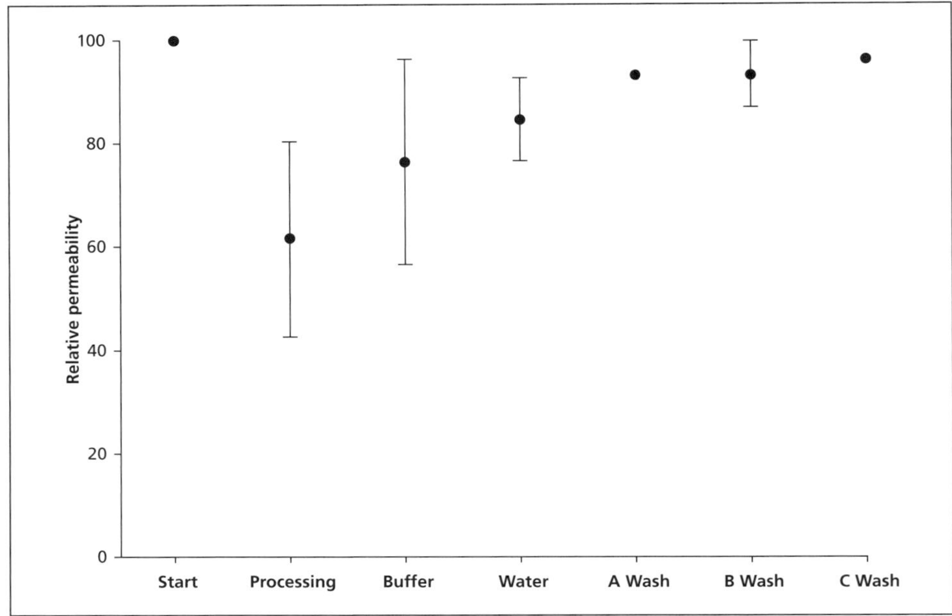

Fig. 36: Cleaning of Planova 20 (n=10-22) filters after operation with biotherapeutics with and without viruses (PPV, MuLV, PI-3, Reo-3) and vendor recommended cleaning. Wash A = 0.1 M NaOH, B = 0.25 M NaOH, C = 0.25 M NaOH, 0.5% Triton X100. Data courtesy of Glen Bolton, Wyeth.

Genentech: Integrity test sensitivity for large assemblies of virus filters *(C. Dowd)*

Virus filter vendors incorporate multiple layers of product and process control in place to ensure the integrity of QC-released virus filters sold on the market. Effective and robust biopharmaceutical virus filtration depends on the control systems of both the filter vendor and the end user. At the point of use, integrity tests (IT) are performed that tie the manufacturing run to the validated virus clearance obtained and confirm the integrity of the virus filter. Typically, gas/water forward flow diffusion integrity tests are employed. However, it could be difficult to detect a single marginal failure in one filter element among many in a large filter assembly. Therefore, Genentech has performed an analysis to estimate the potential virus LRV loss associated with an undetectable hole in one filter element within a larger assembly [46].

Historical manufacturing IT data was used to estimate the worst-case lowest baseline IT gas flow through the filter. It was assumed that one hole was present in one filter within the larger assembly, and that this hole permitted just enough gas flow to meet the integrity test specification. Choked gas flow equations were used to estimate the resulting hole diameter, and the liquid flow through the hole was then calculated. The ratio of liquid flow through the hole to the liquid flow through the remaining integral membrane area was used to estimate an integrity test limited minimum LRV for the assembly as a function of the number of filter elements in an assembly. The integrity test limited minimum LRV is the lowest LRV achievable with a non-integral filter assembly that just meets the integrity test specification.

The largest undetected hole in assembly of twenty 30" filter elements was approximated 140 microns. The integrity test limited minimum LRV was 3-4 $\log_{10}$, and was largely insensitive to the number of filters in the housing. The likelihood of such subtle damage is expected to be very low. Vendor data and information on sterile filter failure rates suggest that such a failure is unlikely to occur in the lifetime of a given biotherapeutic. Although the analysis was initially performed for retrovirus filters, a sensitivity study suggests that very similar results should be expected for parvovirus filters. This analysis supports the integrity testing of large virus filter assemblies as a single unit.

Overall Summary, Virus Filtration

The participants concluded that virus filtration is a reliable and robust technology when run properly. This general conclusion applies to both large and small virus filter types, but clearance of large viruses presents fewer technical challenges. Large virus filters can reliably be expected to clear greater than 6 log_{10} of retroviruses. Presumably, smaller pore size parvovirus filters may also achieve similar clearance of large viruses, but often spike quality issues limit the total clearance that can be measured in scale down studies. When run properly, small virus filters can be expected to clear greater than 3 log_{10} of parvoviruses. However, this application has some potential technological pitfalls. Flow decay can affect clearance of parvovirus across some filters, and in general, overloading should be avoided if there is any evidence of virus breakthrough related to filter fouling or extensive processing. As noted above, the quality of the virus preparation can affect the capacity and measured viral clearance in validation studies of small virus filters. Steps to improve these preparations are described in the PDA Virus Spike Preparation Task Force Technical Report 47 [31]. While size exclusion is thought to be the predominant mechanism of clearance by filters, there may be a subtle pH and buffer effect on parvovirus clearance in small virus filters, probably because of acid induced virus aggregation (see Session VII). Lastly, based on analysis of gas and liquid flow calculations, forward flow integrity testing of multiple virus filter cartridges within a single housing is as effective as testing individual filter elements.

Key areas for process improvement and directed investigation identified at the symposium focused on an assessment of existing data for applicability to modular approaches. For example, some studies in the FDA database reported that retrovirus clearance by small virus retentive filters was not absolute (i.e. ">x log_{10}"). However, did these represent actual breakthrough, as opposed to lab/reporting errors? Due to the retrospective, post-hoc nature of database generation and analysis, the answer remains unclear. A related issue is to what progress can be made towards developing better modular approaches? For example, can generic processing conditions be developed for MMV as a surrogate for a generic claim on retrovirus? (e.g., 4 log_{10}, other?). Can appropriately sized phage be used to evaluate small virus retentive filter performance, at least as surrogates for large viruses (e.g. PR772 at 64-82 nm vs. X-MuLV at 80-110 nm)? Also, these approaches assume that the primary mechanism virus retentive filtration is sieving. The impact of other mechanisms, if they exist, remains to be evaluated.

Other suggestions for process improvement included improved scale-up and validation strategies. For example, cleaner virus preparations as envisioned in PDA TR47 [31] could be developed to minimize clogging during validation studies. Alternatively, custom spiking strategies (e.g. "run-spike", etc.) have been proposed for individual applications [45]. The scalability of filtration is another area for further investigation; issues like membrane non-uniformity could potentially compromise small scale device representativeness. At this point very little data is available in the public domain for discussion of this issue. To proceed further, input from filter vendors to better understand their in-process controls, and testing that determine how much pore size distribution variability exists across membranes, would prove beneficial. Finally, filter re-use was raised as an approach to reduce costs, but the participants felt that this approach raises complex validation issues.

Miesegaes G, Bailey M, Willkommen H, Chen Q, Roush D, Blümel J, Brorson K (eds): Proceedings of the 2009 Viral Clearance Symposium. Dev Biol (Basel). Basel, Karger, 2010, vol 133, p 93.

Closing Remarks and Prioritization of the "Next Steps"

Miesegaes G, Bailey M, Willkommen H, Chen Q, Roush D, Blümel J, Brorson K (eds): Proceedings of the 2009 Viral Clearance Symposium. Dev Biol (Basel). Basel, Karger, 2010, vol 133, p 95.

Closing Remarks and Prioritization of the "Next Steps"

The session and symposium ended with a discussion of conclusions and follow-up items for each unit operation type. These are summarized in the "overall summary" of each unit operations session of this white paper. Many follow-up items were identified, more than can reasonably be expected to be accomplished by a small or medium sized group over a few years. Therefore, the "next steps" were prioritized using a "voice vote" by the participants (Table 38). The prioritization was based on importance, achievability and timeline.

Table 38: Prioritization of proposed viral clearance projects based on conference group consensus.

Unit Op	Next steps	Priority
AEX	Determine pI for viruses (BVDV, Reo3, PRV, Pol, PPV)	High
AEX	Generic clearance studies	High
Low pH	Detailed SOPs for CTOs	High
Low pH	Generic clearance studies	High
Virus filtration	Since no breakthrough of retrovirus through parvofilters... decide whether to just claim MMV values instead, vs. use of phage for parvofilter studies, etc	High
Virus filtration	Characterize/optimize virus preps	High
Detergent only	Impact of lipids on effectiveness of detergent treatment	Med
Low pH	Investigate exceptions noted by regulatory agency	Med
AEX	Investigate possible buffer effects on LRV (e.g. resin specific?)	Low
Protein A	Investigate product vs. resin regarding nonspecific binding of virus; use CHO cells (RVLPs)	Low

The goal of this list is to identify areas of improvement for current industry and regulatory practices for viral clearance by these unit operations. The intention is that conference participants will embark on directed studies to address individual follow-up items identified in the list. While it is probably beyond the resources of one group to follow-up on each item; collaborative efforts are likely to accelerate the process. It is the intention of the conference organizing committee to hold a follow-up meeting to rate progress on each of these items and, to begin the process of defining an integrated virus control strategy for each unit operation and/or the entire process train.

Miesegaes G, Bailey M, Willkommen H, Chen Q, Roush D, Blümel J, Brorson K (eds): Proceedings of the 2009 Viral Clearance Symposium. Dev Biol (Basel). Basel, Karger, 2010, vol 133, p 97.

References

References

1 Kundu A, Reindel K: Evaluation of Viral Clearance in Purification Processes. Biotechnology and Bioprocessing Series 2007;31:419-448.

2 Brorson K: Advances in Viral Clearance, in Shukala A, Etzel M, Gadam S (eds): Process scale bioseparations for the biopharmaceutical industry. Boca Raton, FL, Taylor and Francis, 2007, pp 449-462.

3 Miesegaes G, Lute S, Brorson K: Analysis of Viral Clearance Unit Operations From Monoclonal Antibody Regulatory Submissions. Biotechnol Bioeng 2010; 106(2):319-325.

4 Garnick RL: Raw materials as a source of contamination in large-scale cell culture, in Brown F, Griffiths E, Horaud F, Petricciani JC (eds): Safety of Biological Products Prepared from Mammalian Cell Culture. Dev Biol Stand. Basel, Karger, 1998, vol 93, pp 21-29.

5 Lieber MM, Benveniste RE, Livingston DM, Todaro GJ: Mammalian cells in culture frequently release type C viruses. Science 1973;182(107):56-59.

6 Xu Y, Brorson K: An overview of quantitative PCR assays for biologicals: quality and safety evaluation, in Brown F, Lubiniecki AS (eds): Process Validation for Manufacturing of Biologics and Biotechnology Products. Dev Biol, Basel, Karger, 2003, vol 113, pp 89-98.

7 Lute S, Norling L, Hanson M, Emery R, Stinson D, Padua K, Blank G, Chen Q, Brorson K: Robustness of virus removal by Protein A chromatography is independent of media lifetime. J Chromatogr A 2008;1205(1-2):17-25.

8 Zhang M, Lute S, Norling L, Hong C, Safta A, O'Connor D, Bernstein LJ, Wang H, Blank G, Brorson K, Chen Q: A novel, Q-PCR based approach to measuring endogenous retroviral clearance by capture protein A chromatography. Biotechnol Bioeng 2009;102(5):1238-1247.

9 Brorson K, Krejci S, Lee K, Hamilton E, Stein K, Xu Y: Bracketed generic inactivation of rodent retroviruses by low pH treatment for monoclonal antibodies and recombinant proteins. Biotechnol Bioeng 2003;82(3):321-329.

10 Boschetti N, Niederhauser I, Kempf C, Stuhler A, Lower J, Blumel J: Different susceptibility of B19 virus and mice minute virus to low pH treatment. Transfusion 2004;44(7):1079-1086.

11 Norling L, Lute S, Emery R, Khuu W, Voisard M, Xu Y, Chen Q, Blank G, Brorson K: Impact of multiple re-use of anion-exchange chromatography media on virus removal. J Chromatogr A 2005;1069(1):79-89.

12 Booth J, Vicik S, Tannatt M, Gallo C, Kelley B: Transmissible spongiform encephalopathy agent clearance by the immunoaffinity and anion-exchange chromatography steps of the ReFacto manufacturing process. Haemophilia 2007;13(5):580-587.

13 Ishihara T, Kadoya T: Accelerated purification process development of monoclonal antibodies for shortening time to clinic. Design and case study of chromatography processes. J Chromatogr A 2007;1176(1-2):149-156.

14 Knudsen HL, Fahrner RL, Xu Y, Norling LA, Blank GS: Membrane ion-exchange chromatography for process-scale antibody purification. J Chromatogr A 2001;907(1-2):145-154.

15 Phillips M, Cormier J, Ferrence J, Dowd C, Kiss R, Lutz H, Carter J: Performance of a membrane adsorber for trace impurity removal in biotechnology manufacturing. J Chromatogr A 2005;1078(1-2):74-82.

16 Zhou JX, Tressel T: Basic concepts in Q membrane chromatography for large-scale antibody production. Biotechnol Prog 2006;22(2):341-349.

17 Zhou JX, Tressel T, Gottschalk U, Solamo F, Pastor A, Dermawan S, Hong T, Reif O, Mora J, Hutchison F, Murphy M: New Q membrane scale-down model for process-scale antibody purification. J Chromatogr A 2006;1134(1-2):66-73.

18 Follman DK, Fahrner RL: Factorial screening of antibody purification processes using three chromatography steps without Protein A. J Chromatogr A 2004;1024(1-2):79-85.

19 Strauss DM, Lute S, Tebaykina Z, Frey DD, Ho C, Blank GS, Brorson K, Chen Q, Yang B: Understanding the mechanism of virus removal by Q sepharose fast flow chromatography during the purification of CHO-cell derived biotherapeutics. Biotechnol Bioeng 2009;104(2):371-380.

20 ICH, Viral safety evaluation of biotechnology products derived from cell lines of human or animal origin, Q5A, 1999; International Conference on Harmonisation of Technical Requirements for Registration of Pharmaceuticals for Human Use; Geneva, Switzerland.

21 Riordan W, Heilmann S, Brorson K, Seshadri K, He Y, Etzel M: Design of salt-tolerant membrane adsorbers for viral clearance. Biotechnol Bioeng 2009;103(5):920-929.

22 Curtis S, Lee K, Blank GS, Brorson K, Xu Y: Generic/matrix evaluation of SV40 clearance by anion exchange chromatography in flow-through mode. Biotechnol Bioeng 2003;84(2):179-186.

23 Strauss DM, Gorrell J, Plancarte M, Blank GS, Chen Q, Yang B: Anion exchange chromatography provides a robust, predictable process to ensure viral safety of biotechnology products. Biotechnol Bioeng 2009;102(1):168-175.

24 Robertson J, Blumel J, K, B, Groner A, Kreil T, Ruiz S, Willkommen H: Meeting Report - Virus and TSE Safety Forum. Biologicals 2009;37(5):345-354.

25 Strauss DM, Lute S, Brorson K, Blank GS, Chen Q, Yang B: Removal of endogenous retrovirus-like particles from CHO-cell derived products using Q sepharose fast flow chromatography. Biotechnol Prog 2009;25(4):1194-1197.

26 Brorson K, Shen, H, Lute S, Perez J.S, Frey DD: Characterization and purification of bacteriophages using chromatofocusing. J Chromatogr A 2008;1207(1-2):110-121.

27 Dichtelmuller HO, Biesert L, Fabbrizzi F. Gajardo R, Groner A, von Hoegen I, Jorquera JI, Kempf C, Kreil TR, Pifat D, Osheroff W, Poelsler G: Robustness of solvent/detergent treatment of plasma derivatives: a data collection from Plasma Protein Therapeutics Association member companies. Transfusion 2009;49(9):1931-1943.

28 Horowitz B, Piet MP, Prince AM, Edwards CA, Lippin A, Walakovits LA: Inactivation of lipid-enveloped viruses in labile blood derivatives by unsaturated fatty acids. Vox Sang 1988;54(1):14-20.

29 Prince AM, Stephan W. Dichtelmuller H, Brotman B, Huima T: Inactivation of the Hutchinson strain of non-A, non-B hepatitis virus by combined use of beta-propiolactone and ultraviolet irradiation. J Med Virol 1985;16(2):119-125.

30 Roberts P: Resistance of vaccinia virus to inactivation by solvent/detergent treatment of blood products. Biologicals 2000;28(1):29-32.

31 PDA, Technical report 47: Preparation of virus spikes used for virus clearance studies. Parenteral Drug Association, Bethesda, MD, 2010.

32 FDA, Points to Consider in the Manufacture and Testing of Monoclonal Antibody Products for Human Use, 1997; Department of Health and Human Services, Food and Drug Administration, Rockville, MD.

33 Biesert L, Suhartono H: Solvent/detergent treatment of human plasma--a very robust method for virus inactivation. Validated virus safety of OCTAPLAS. Vox Sang 1998;74(S1):207-212.

34 Horowitz B, Lazo A, Grossberg H: Virus inactivation by solvent/detergent treatment and the manufacture of SD-plasma. Vox Sang 1998;74(S1):203.

35 Miesegaes G, Lute S, Aranha H, Brorson K: Virus Retentive Filtration, in Flickinger M (ed): Encyclopedia of Industrial Biotechnology. Hoboken NJ, Wiley Interscience, 2010.

36 PDA, Technical report 41: Virus retentive filtration. Parenteral Drug Association, Bethesda, MD, 2008.

37 Lute S, Bailey M, Combs J, Sukumar M, Brorson K: Phage passage after extended processing in small-virus-retentive filters. Biotechnol Appl Biochem 2007;47(Pt 3):141-151.

38 Bolton G, Cabatingan M, Rubino M, Lute S, Brorson K, Bailey M: Normal-flow virus filtration: detection and assessment of the endpoint in bio-processing. Biotechnol Appl Biochem 2005;42(Pt 2):133-142.

39 Levy R, Brorson K: Consensus rating methods for small and large virus-retentive filters. Am Pharm Rev 2009;Mar/Apr:69-73.

40 Brorson K, Sofer G, Robertson G, Lute S, Martin J, Aranha H, Haque M, Satoh S, Yoshinari K, Moroe I, Morgan M, Yamaguchi F, Carter J, Krishnan M, Stefanyk J, Etzel M, Riorden W, Korneyeva M, Sundaram S, Wilkommen H, Wojciechowski P: "Large pore size" virus filter test method recommended by the PDA Virus Filter Task Force. PDA J Pharm Sci Technol 2005;59(3):177-186.

41 Grant D, Liu B, Fisher W, Bowling R: Particle Capture Mechanisms in Gases and Liquids: An Analysis of Operative Mechanisms in Membrane/Fibrous Filters. The Journal of Environmental Sciences 1989;32(4):43-51.

42 Cabatingan M: Impact of virus stock quality on virus filter validation: A case study. Bioprocess International 2005;3(10):S:39-43.

43 Zhou JX, Solamo F, Hong T, Shearer M, Tressel T: Viral clearance using disposable systems in monoclonal antibody commercial downstream processing. Biotechnol Bioeng 2008;100(3):488-496.

44 Romanowski P, Schleh M, Rajurs V, Chinniah S, Dehghani H: Variables Affecting Titer and Long-Term Stability of Virus Stocks: Xenotropic Murine Leukemia and Mouse Minute Viruses. Bioprocess International 2008;6(2):44-53.

45 Khan NZ, Parrella JJ, Genest PW, Colman MS: Filter preconditioning enables representative scaled-down modelling of filter capacity and viral clearance by mitigating the impact of virus spike impurities. Biotechnol Appl Biochem 2009;52(Pt 4):293-301.

46 Dowd CJ: Multi-round virus filter integrity test sensitivity. Biotechnol Bioeng 2009;103(3):574-581.

Developments in Biological Standardization

Vol. 95	Paris, 1997 (296 p.)	Preclinical and Clinical Development of New Vaccines
Vol. 96	Washington, DC, 1995 (224 p.)	Characterization of Biotechnology Pharmaceutical Products
Vol. 97	London, 1997 (204 p.)	Biological Characterization and Assay of Cytokines and Growth Factors
Vol. 98	London, 1997 (224 p.)	Inactivated Influenza Vaccines Prepared in Cell Culture
Vol. 99	Strasbourg, 1998 (216 p.)	Animal Sera, Animal Sera Derivatives and Substitutes Used in the Manufacture of Pharmaceuticals: Viral Safety and Regulatory Aspects
Vol. 100	Geneva, 1998 (192 p.)	A Celebration of 50 Years of Progress in Biological Standardization and Control at WHO
Vol. 101	London, 1998 (352 p.)	Alternatives to Animals in the Development and Control of Biological Products for Human and Veterinary Use

Developments in Biologicals

Vol. 102	S. Francisco, CA, 1999 (264 p.)	Advances in Transfusion Safety
Vol. 103	Annecy, 1999 (292 p.)	Physico-Chemical Procedures for the Characterization of Vaccines,
Vol. 104	Langen, 1999 (224 p.)	Development and Clinical Progress of DNA Vaccines
Vol. 105	Paris, 2000 (260 p.)	Progress in Polio Eradication: Vaccine Strategies for the End Game
Vol. 106	Rockville, MD, 2000 (560 p.)	Evolving Scientific and Regulatory Perspectives on Cell Substrates for Vaccine Development
Vol. 107	London, 2000 (152 p.)	The Design and Analysis of Potency Assays for Biotechnology Products
Vol. 108	Langen, 2001 (160 p.)	Advances in Transfusion Safety
Vol. 109	Washington, DC, 2000 (168 p.)	Biologics 2000 – Comparability of Biotechnology Products
Vol. 110	Palm Cove, 2001(192 p.)	Orphan Vaccines – Bridging the Gap
Vol. 111	Utrecht, 2001 (356 p.)	Advancing Science and Elimination of the Use of Laboratory Animals for Development and Control of Vaccines and Hormones
Vol. 112	Bethesda, MD, 2001 (188 p.)	Immunogenicity of Therapeutic Biological Products
Vol. 113	Berlin, 2001 (144 p.)	Process Validation for Manufacturing of Biologics and Biotechnology Products

Vol. 114	Ames, IA, 2002 (312 p.)	Vaccines for OIE List A and Emerging Animal Diseases
Vol. 115	Bergen, 2002 (176 p.)	Laboratory Correlates of Immunity to Influenza – A Reassessment
Vol. 116	Los Angeles, 2003 (260 p.)	Development of Therapeutic Cancer Vaccines –
Vol. 117	Langen, 2003 (192 p.)	Consideration of Alternative Licensing Procedures for Vaccines for Minor Species, Minor Indications and Autogenous/ Autologus Products
Vol. 118	Langen, 2003 (192 p.)	PDA/EMEA European Virus Safety Forum
Vol. 119	Buenos Aires, 2004 (528 p.)	Control of Infectious Animal Diseases by Vaccination
Vol. 120	Bethesda, 2003 (240 p.)	Advances in Transfusion Safety
Vol. 121	Bergen, 2003 (352 p.)	Progress in Fish Vaccinology
Vol. 122	Bethesda, 2003 (224 p.)	State of the Art Analytical Methods for Characterization of Biological Products and Assessment of Comparability
Vol. 123	Rockville, 2004 (384 p.)	Vaccine Cell Substrates 2004
Vol. 124	Paris, 2005 (288 p.)	International Scientific Conference on Avian Influenza
Vol. 125	Kiev, 2005 (344 p.)	Rabies in Europe
Vol. 126	St. Malo, 2005, (352 p.)	New Diagnostic Technology: Applications in Animal Health & Biologics Controls,
Vol. 127	Sydney, 2005 (288 p.)	Advances in Transfusion Safety – Volume IV
Vol. 128	Florianopolis, 2006 (216 p.)	First International Conference of the OIE Reference Laboratories and Collaborating Centres
Vol. 129	Bergen, 2006 (204 p.)	The OIE Global Conference on Aquatic Animal Health
Vol. 130	Verona, 2007 (208 p.)	Vaccination: A Tool for the Control of Avian Influenza
Vol. 131	Paris, 2007 (608 p.)	Towards the Elimination of Rabies in Eurasia
Vol. 132	Paris, 2007 (456 p.)	Animal Genomics for Animal Health